ILL WINDS:
Airborne Pollution's Toll on Trees and Crops

James J. MacKenzie
Mohamed T. El-Ashry

WORLD RESOURCES INSTITUTE
A Center for Policy Research

September 1988

Kathleen Courrier
Publications Director

Don Strandberg
Marketing Manager

Hyacinth Billings
Production Supervisor

James J. MacKenzie (cars, trees)
Tennessee Valley Authority (power plant)
U.S. Department of Agriculture (wheat)
Cover Photos

Each World Resources Institute Report represents a timely, scientific treatment of a subject of public concern. WRI takes responsibility for choosing the study topics and guaranteeing its authors and researchers freedom of inquiry. It also solicits and responds to the guidance of advisory panels and expert reviewers. Unless otherwise stated, however, all the interpretation and findings set forth in WRI publications are those of the authors.

Copyright © 1988 World Resources Institute. All rights reserved.
Library of Congress Catalog Card Number 88-051128
ISBN 0-915825-29-5

Contents

I. Introduction 1

II. Damage to Forests and Crops 3
   Forest Damage 3
   Crop Damage 9

III. Links Between Air Pollution and Damage
   to Forests 13
   Air Pollution's Impacts on California's
      Forests............................13
   Air Pollution and Forest Declines in the
      Eastern United States 14

IV. Links Between Air Pollution and Damage
   to Crops..............................25
   Identifying Ozone's Impacts on
      Crops..............................25
   Quantifying the Impacts 26
   Estimating Productivity Losses........31

V. Air Pollution in the U.S.: Sources,
   Trends, and Proximity to Forest and
   Agricultural Damages..................33

Trends in Pollutant Emissions 33
Formation of Secondary Pollutants 41
Regional Patterns of Pollutant
   Concentrations and Deposition 44
Patterns of Emissions and Patterns of
   Damages...........................48

VI. Summary and Policy Recommendations ..51
   Summary of Pollution Damages to
      Forests and Crops 51
   Economic Impacts of Air Pollutants 52
   Future Pollution Trends...............53
   Structural Shortcomings of the Clean
      Air Act 54
   Air Pollution and Other National
      Issues 55
   An Integrated Approach to Pollution
      Control 56

Notes....................................61

Appendix 73

Acknowledgments

In the course of this study we have consulted with many colleagues from Europe, Canada, and the United States. We are especially grateful to the authors of the background papers we commissioned and to the members of our Advisory Panel, who provided helpful guidance on the design of the study and constructive comments on early drafts. Their names and affiliations are listed in the appendix. While not all contributing authors or members of the Advisory Panel may agree with all of our conclusions and recommendations, all their comments and suggestions were constructive and helpful. Ultimately, of course, the authors alone are responsible for the accuracy and recommendations of the report.

Our thanks to Jessica Mathews, Robert Repetto, and Bill Moomaw for their comments on the draft report, to Kathleen Courrier for her patience and skill in editing the report, to Hyacinth Billings and Allyn Massey for their assistance in preparing the text, figures, and cover, and to Beverlyann Austin, Laura Lee Dooley, Donna Pirlot, and Maggie Powell for their assistance with the manuscript. Thanks also to Michael P. Walsh and Susan Garbini for their reports on transportation and clean coal technologies, respectively, and to Chris Bernabo for many helpful discussions on forest decline. Our special thanks to Robert I. Bruck, Arthur H. Johnson, Robert A. Gregory, Reinhard F. Huettl, and Peter Schuett for accompanying us on field trips to various forest research sites in the Federal Republic of Germany and North America.

Finally, our gratitude goes to Gus Speth for his overall advice and guidance.

JJM
MTE

Foreword

Our century has witnessed unprecedented growth in human population and economic activity. World population has increased more than threefold; gross world product by perhaps twentyfold. An engine of this growth, fossil fuel use has climbed more than tenfold. With these huge increases in economic activity and fossil fuel use have come dramatic changes in the quantities of pollutants released. Between 1900 and 1985, annual sulfur dioxide emissions increased sixfold globally, while nitrogen oxide emissions increased about tenfold, perhaps more. These gases, together with hydrocarbons, are principal sources of both urban air pollution and acid rain.

Fossil fuel combustion also forms carbon dioxide, one of the greenhouse gases implicated in global warming and climate change. Annual global emissions of CO_2 have increased tenfold in this century, and a dramatic 25-percent increase in the CO_2 content of Earth's atmosphere has occurred.

When pollution volumes were much smaller, most of their impacts were confined to limited areas near sources. As late as 1970, it was possible to write national clean air legislation in the United States that focussed almost exclusively on local pollution. Today, the scale and intensity of air pollution make its consequences both regional and global.

Since the early 1980s, some 23,000 square miles of Central Europe's forests have been plagued by a death-dealing combination of airborne pollutants and natural stresses. The result, *neuartige Waldschaeden* ("new type of forest damage"), has alarmed the forestry industry, scientists, and the European public—initially because the symptoms didn't match any in the technical literature on forest diebacks and more recently because the toll is rising.

Considering that U.S. tree species and pollution levels resemble Europe's, is the United States next in line for this scourge? Investigating recent damages to trees and to agricultural crops in the United States, WRI's James J. MacKenzie and Mohamed T. El-Ashry have assembled disturbing evidence that it may be. California pines are declining; red spruce are dying all along the Appalachian chain; growth reduction is occurring in eastern white pine; and maple and other hardwoods in northern New England are growing irregularly and perishing prematurely.

As the review of scientific evidence in this report shows, ozone pollution is to blame for the death of pine stands in California and for the losses among sensitive eastern white pines, as well as for billions of dollars of losses in U.S. crop productivity. A life-saving shield in the stratosphere, ozone at ground level stunts vegetative growth. There is growing evidence that both ozone and acid deposition are important contributors to the abnormally high rates of mortality among spruce trees along the Appalachians from Vermont to North Carolina.

Air pollution is also a leading suspect in the growth decline of yellow pines in the Southeast.

As *Ill Winds* makes all too clear, ozone and acid deposition, singly or combined, are contributing to forest decline in many ways that vary by site. This complexity makes it difficult to establish beyond dispute exactly what causes damage at each site. But current evidence strongly suggests that air pollution is an important element in observed declines. No other explanation ventured so far can explain the injury of so many species over so much territory.

Like many other environmental menaces, losses of crops and trees will challenge policymakers. The causes are still being debated; understanding of the long-term impacts awaits further research and monitoring; and the biological and economic benefits of reducing emissions aren't nearly as clearcut as the costs. Yet, just as *Ill Winds* underscores the long-term threats to U.S. forests and crops, it also connects sulfur oxides, nitrogen oxides, and hydrocarbons to a host of other maladies: human illness, acidified lakes and streams, degraded visibility, and harm to buildings and other structures. These pollutants, and the bulk of the carbon dioxide contributing to the greenhouse effect and global warming, are all products of our heavy and extravagant use of fossil fuels in transportation, power plants, and industry.

The policy recommendations in *Ill Winds* call for additional pollution control, but they also stress the need for more efficient and environmentally benign U.S. energy and transportation policies. The authors' proposals would reduce emissions from both power plants and vehicles for the next ten to twenty years by making today's technology cleaner and more efficient while encouraging a more gradual shift to non-fossil energy sources.

WRI would like to express its great appreciation to the Joyce Foundation, the Geraldine R. Dodge Foundation, and the John D. and Catherine T. MacArthur Foundation, whose financial support made this research possible.

James Gustave Speth
President
World Resources Institute

I. Introduction

Widespread damage to trees and crops is not a late-1980s phenomenon. Such natural causes as drought, unseasonal heat or cold, high winds, diseases, and insects have long been known to injure and kill both.

Scientists have also recognized for many decades that air pollution can injure vegetation.[1] Historically, smelters, power plants, and other large "point sources" of pollution have acutely damaged vegetation downwind; usually, high levels of sulfur dioxide or fluoride (generally as hydrogen fluoride) were at fault. Within the past thirty years, such photochemical oxidants as ozone have also been shown to cause vegetation damage.

For ten to twenty-five years now, an unusually large number of dead and damaged trees have been observed in Central Europe and the United States, though at first no obvious causes were identified. Visible foliar symptoms first appeared in Germany in the late 1970s and by 1980 had spread to many European countries. In Europe, the symptoms developed more rapidly than in the United States, and now almost all important species at all elevations are in decline.[2]

In the United States, tree injury and crop damage is occurring across the entire nation, from California to Maine. Tree mortality has been most extensive in California and high in the Appalachian Mountains from North Carolina up through New England; costly crop injury is occurring through much of the U.S. breadbasket.

California's losses, mostly pines, began in the late 1940s; Appalachia's, mostly spruce, began showing symptoms of decline as early as 1960. Growth reductions among low-elevation red spruce in the East, among commercial yellow pines in the Southeast, and in New Jersey's Pine Barrens have also been observed. Growth reductions and damages have been documented for sensitive eastern white pines too. And sugar maples in the Northeast and Canada are losing their crowns and dying prematurely.

Most declines probably reflect multiple stresses acting together, rather than a comparatively easy-to-discover single cause.

The causes of these damages can seem baffling. Forests are complex ecosystems, acted upon by nature and man alike. Most declines probably reflect multiple stresses acting together, rather than a comparatively easy-to-discover single cause. Many contributing causes may lead to decline *only* collectively.

In Europe, scientists generally agree that air pollutants number among the primary causes of forest decline.[3] No other factor can explain the near-synchronous decline of so many different species over so vast an area. As a result, considerable research has been devoted to understanding the possible role of multiple air pollutants (especially acid deposition and ozone).

In the United States, a somewhat similar research program (the National Acid Precipitation Assessment Program), under way since 1982, was launched to identify the causes and effects of acid deposition and related pollutants. In some U.S. declines, the principal causes are already well known. Losses of ponderosa and Jeffrey pines in southern California and of eastern white pines are two cases in point. In both of these declines, ozone has been shown to contribute to tree damage and to increase susceptibility to mortality from such other natural factors as drought, insects, and weather extremes. In contrast, there is still scientific uncertainty over the declines of red spruce and Fraser fir at high elevations in the eastern United States and of the sugar maple in the Northeast, though evidence summarized in this report indicates that—at least in the case of the fir-spruce decline—air pollution is an important factor. *(See Section III.)*

How air pollution affects agricultural crops was the question asked in a major EPA-sponsored research program begun in 1980, the National Crop Loss Assessment Network (NCLAN). This recently completed research demonstrated that ozone is the principal air pollutant reducing crop productivity. At current concentrations, ozone is causing annual losses in excess of $3 billion for major crops alone.

In this report, evidence that air pollutants are contributing to current forest and agricultural losses in the United States is presented and policy options to reduce these losses explored. The analysis presented here has two special features. First, discussion and analysis are not confined to a single pollutant. Examined instead is all the evidence that multiple pollutants—especially such secondary pollutants as acidic compounds and ozone—are contributing to the problem, both separately and together. Second, policy recommendations are made within a broad framework of both short- and long-term national energy planning, *not* in terms of narrower efforts to meet ambient air standards by regulating individual sources of pollution. This latter approach simply fails to take into account growth in the number of pollution sources, as well as the long-range transport and interactions among pollutants. In contrast, the recommendations made here do take into account such other critical national issues as national energy security, failure to attain clean air standards in our major cities, and climate protection.

II. Damage to Forests and Crops

Many influences shape the overall health and growth of trees and crops. Some of these influences are natural—including competition among species, changes in precipitation, temperature fluctuations, insects, and disease. Others result from air pollution, use of pesticides and herbicides, logging, land-use practices, and other human activities. With so many possible stresses at play, determining precisely which are to blame when trees die in large numbers or crop yields fall is difficult indeed.[4] But the research summarized later in this report strongly implicates air pollution as an important factor.

Crop failures are usually easier to diagnose than are widespread tree declines. By nature, agricultural systems tend to be more highly managed and ecologically simpler than forests. Also, much larger resources have been devoted to developing and understanding agricultural systems than to understanding natural forests. Relative to agricultural science, forest ecology is still in its infancy, and many years of painstaking field research may be required to fully grasp the etiology of specific forest declines.

Historically, observed injuries to forest trees and crops have prompted scientific research to uncover the causes. In some cases, such as pine damage and mortality in southern California, the principal initiating agent was eventually identified, ozone air pollution. In others—including the ongoing decline of red spruce and Fraser fir in the eastern United States—intense research is still under way. (Even with the spruce-fir decline, though, evidence indicates that air pollution is an important predisposing factor.)

In the following review of evidence that injuries to forest trees and crops are occurring in the United States, only those examples where researchers have either demonstrated that pollution is a contributing factor or are actively investigating such a possibility are included. Damage considered to be the result of such purely natural factors as drought, disease, or insects is deliberately excluded. The story that unfolds is that significant plant injury is occurring across the country. In the West, sensitive pines are dying; throughout the Midwest, crop productivity is being impaired; and in the East, spruce, fir, maple, and white and yellow pines are either dying outright or growing abnormally slowly. In short, from one coast to the other, significant vegetative injury is occurring and much of it has already been shown to result from air pollution. *(For a review of the scientific evidence linking the symptoms described below with various forms of air pollution, see Sections III and IV.)*

Forest Damage

As popularly used, the term "decline" describes the death of large numbers of trees. In forestry, however, "decline" has a fairly specific meaning. According to forestry expert Paul D. Manion, "decline refers to those situations in which a deterioration and death of trees is caused by a combination of biological and non-biological stress factors."[5] Indeed, the combination of contributing factors is an important part of the concept of decline. Chestnut blight and Dutch elm disease, for example, have single causes, so they are not considered declines, even though both have killed many trees.

Declines are best understood as what happens when natural and anthropogenic stresses acting together weaken and eventually kill trees. Manion has organized these stresses into three consecutive categories: predisposing, inciting, and contributing factors.[6] Predisposing factors include poor soils, climate change, air pollution, and the genetic makeup of the trees.[7] Inciting factors—including insects, drought, and frost—weaken predisposed trees until they succumb to contributing lethal factors. Finally, contributing factors—viruses, root rot, insects, and fungi, among them—may finish the trees off.

In two such interplays of stresses, the ash declines of the 1930s and the pole blight of Western white pine in the 1940s, drought was the initiating factor.[8] Once the trees were weakened, disease killed great numbers of trees. In earlier sugar maple declines, contributing stresses include drought, soil compaction, over-tapping, and insect defoliation.[9] In these declines, no single factor explains the dieback.

Frequently, the trained forester, forest pathologist, or forest entomologist can determine the causes of many kinds of tree damage.[10] Injury caused by insects can usually be recognized by the general appearance of the tree and by evidence of past or present insect activity.[11] Damage caused by wind, snow, fire, drought, and some damages caused directly by air pollutants are also readily identified. But in some recent declines, even experts cannot diagnose the cause of specific injuries.[12]

Numerous forest declines have hit both Europe and North America over the past 100 to 200 years. At least five regional declines have occurred in Europe, and at least 13 others have taken place in North America in this century alone.[13] The symptoms in these declines have varied widely, and some appear in just one tree species.[14] Air-borne chemicals are considered important (or potentially so) in six of these episodes: the decline of pines in the San Bernardino Mountains of California; white pine mortality in the eastern United States; Europe's *Waldschaeden* (described below); the red spruce and Fraser fir declines in the eastern United States; the decline in growth of yellow pines in the southeastern United States; and the recent symptomatically distinct decline of sugar maples in the northeastern United States.[15] Natural factors—disease, insects, competition, weather extremes, and so forth—were the principal causes of the remaining declines and, indeed, may well have played a role in the others.

Historically, high levels of air pollution have contributed to the death of trees near large power plants and other industrial sources.[16] In the early 20th century, a copper smelter in Trail, British Columbia, that emitted more than 10,000 tons of sulfur dioxide per month killed or severely damaged more than 30 percent of the trees for 52 miles south of the plant, and 60 percent of the trees for 33 miles south of the plant.[17] In the early 1950s in Spokane, Washington, an aluminum ore reduction plant's fluoride emissions killed all the trees within a three square-mile area near the plant.[18] Significant foliar damage was observed in a 50-square mile area.

Over the past three decades, a number of forest declines observed in the western United States (especially California), in central Europe, and along the Appalachian Mountains from North Carolina to New Hampshire and Vermont appear to differ significantly from those of the past.[19] These declines and their curious symptoms stimulated scientists' current concern over forest health.

Pines in San Bernardino. The San Bernardino forest, some 75 miles east of Los Angeles, is a predominantly mixed conifer forest of ponderosa pine, Jeffrey pine, sugar pine, white fir, and incense cedar.[20] Damage to ponderosa pine trees in the San Bernardino National Forest was first noted in the early 1950s. By 1962, an estimated 25,000 acres of the mixed-conifer forest had been injured. Subsequent surveys (1969) showed that ponderosa and Jeffrey pines on 100,000 to 160,000 acres of forest were moderately to severely damaged.[21]

The symptoms observed in the San Bernardino forest were described as chlorosis (yellowing) of older needles, leading to premature senescence. The problem was dubbed "x-disease" because its cause eluded scientists at the time. Although 1946 and 1953 were the driest years then on record, the symptoms did not match those caused by drought, insects, or disease. What's more, ostensibly healthy pines and other more drought-sensitive species were growing alongside diseased trees. The damaged pines showed decreased radial growth and reduced tolerance to the western pine beetle and other stresses. Later surveys revealed similar damage in the Laguna Mountains east of San Diego, and in the Sierra Nevada and San Gabriel Mountains.[22] In 1983, about one fourth of the trees on established plots in the Sequoia National Forest showed a yellow mottle on new needles, compared with 14 percent in 1975. More recent surveys have shown that ponderosa and Jeffrey pines are suffering increasing damage over time. While 48 percent of all surveyed trees showed some injury in 1980–82, 87 percent did by 1985.[23] Later research established that ozone is the principal cause of "x-disease" and that ponderosa and Jeffrey pine are particularly sensitive to this pollutant.[24]

White Pines in the Eastern United States. White pine trees grow widely throughout southeastern Canada and the eastern United States. In New England, white pines grow from sea level up to 500 meters in elevation; in the southern Appalachians, they can be found up to 1200 meters.[25]

Within the past few decades researchers have found that sensitive white pines are being damaged by ozone over much of the eastern United States.[26] Most of the injury occurs during June and July, the high-ozone season. Visible foliar injury begins with needle flecking and proceeds until necrotic bands develop and the needle tip dies. Sensitive white pines in high-ozone regions grow less in height, diameter, and needle length than more resistant genotypes.[27] Over the long-term, some researchers believe, ozone exposures will eliminate sensitive genotypes of the species.[28]

Forest Declines in Central Europe. The multi-species forest decline that central Europe is now witnessing was first recognized on low mountains in the late 1970s.[29] Silver fir (*Abies alba*) in the Federal Republic of Germany's Black Forest began to display unusual symptoms of disease—loss of needles from the inside of the branches outward and from the bottom upward to the crown—and to die in large numbers.[30] Soon afterwards, Norway spruce (*Picea abies*) began to show similar symptoms: chlorosis of the needles (clearly linked with nutritional deficiencies, especially of magnesium, calcium, zinc, potassium, and manganese[31]), and defoliation of the older trees. Pine was next to show symptoms (thinned crowns), followed in 1982 by the hardwoods—mainly beech and oak. The crown leaves of beech trees were yellowing and dropping off in early summer, and leaves and branches were forming abnormally. Discoloration and premature leaf fall have now been reported for essentially all major forest tree species in West Germany.[32]

Although damage in West Germany first appeared at elevations of 800 meters and higher, it has now spread to lower-lying areas. Damage is worst on west-facing slopes, which face prevailing winds. Although trees of all ages are affected, older trees show the most symptoms.

Responding to the increasing damage observed, West Germany began forest surveys in 1982. By 1984, the surveys had become standardized and applied uniformly in all eleven

German states. The 1983 survey showed that 34 percent of West Germany's trees were affected. In 1984, 1985, and 1986, the percentages climbed to 50, 52, and 54, respectively, though only about one in five of these damaged trees lost more than 25 percent of its leaves.[33] *(See Table 1.)* Fir trees showed the heaviest damage: 83 percent exhibited some symptoms and over 60 percent showed moderate to severe damage. Injury to oak, beech, and spruce has also been substantial.

Symptoms of *neuartige Waldschaeden*—literally, new type of forest damage—have now been detected over most of central Europe on all types of soils and at all elevations. (*Waldschaeden* has generally replaced the older term *Waldsterben*—forest death—as a description of the European problem.) By 1983, injury was detected in Switzerland, with about 14 percent of all Swiss forests showing symptoms of decline;[34] by 1984, 34 percent were damaged. A 1985 forest survey in Austria found that 22 percent of forests were damaged, 4 percent moderately to heavily. In France, according to a 1984 survey, 22 percent of the coniferous trees and 4 percent of the deciduous trees displayed pronounced damage. In Holland, a 1986 survey showed that 29 percent of forest areas were marked by moderate to heavy symptoms of decline.

In a recent review of the data on the decline in Europe's coniferous and deciduous forests, researchers found that 15 percent of Europe's coniferous forests (measured by tree volume) are dead or moderately or severely damaged.[35] In the United Kingdom and Sweden, 20 percent of all conifers fell within one of these categories, and more than 17 percent of deciduous trees (by volume) are damaged.

The Norway spruce, central Europe's most common and most important commercial tree species, is also one of the hardest hit. Studied extensively, the symptoms of decline shown by Norway spruce are site-specific and include yellowing of needles at high elevations, crown thinning at middle elevations and in coastal areas, and needle necrosis in older stands.[36]

High-elevation yellowing, a new phenomenon, occurs predominantly on older needles

Table 1. 1986 Forest Damage in West Germany by Species and Damage Class (in Percent)

Damage Stage	Spruce	Pine	Fir	Beech	Oak	Others	Total
Undamaged	45.9	46.0	17.1	39.9	39.3	65.8	46.3
Slightly Damaged	32.4	39.5	22.5	41.2	41.2	24.5	34.8
Moderately Damaged	20.1	13.1	49.1	17.5	18.7	8.5	17.3
Seriously Damaged or Dead	1.6	1.4	11.4	1.4	0.8	1.2	1.6
All Damage Classes	54.1	54.0	82.9	60.1	60.7	34.2	53.7

Source: Waldschadenserhebung 1986; Bundesministerium fuer Ernaehrung, Landwirtschaft and Forsten, Table 5.

and twigs directly exposed to sunlight. Yellowed trees also have below-normal radial growth rates. Such yellowing has been observed in Austria, France, Belgium, and the Netherlands.

Crown thinning at middle elevations is widespread, mainly on nutrient-poor sites at 400 to 600 meters and in forests where acid deposition reaches relatively high levels. Needle loss may or may not be preceded by yellowing, and usually the older needles fall first.

Curiously, moderate needle loss (up to 40 or 50 percent) in older foliage can occur among Norway spruce without significant loss of radial growth. Norway spruce in the Bavarian Alps, observed to be heavily damaged in 1975, were shown to have experienced increased growth in the previous 25 years.[37] In the case of silver fir, on the other hand, growth falls off significantly when needle losses exceed 25 to 30 percent.

Needle necrosis in spruce stands older than 60 years begins as an orange-yellow discoloration among older needles. After two weeks, the needles turn reddish brown and fall off. The crowns become progressively thinner, though the trees seldom die as a result. Younger needles appear healthy, and no long-term growth suppression has been observed.[38]

Significant spruce declines characterized by needle discoloration and loss have also been observed on steep slopes and ridges in the calcareous Alps, where the carbonate-derived soils are shallow. The overpopulation of wild game species on these sites impedes natural regeneration, so grassy gaps without trees invite soil erosion from rain and snow and endanger the whole ecosystem.[39] Needle discoloration has long been associated with deficiencies of potassium and manganese, but the past few years have seen an unusual increase in yellowing, along with a loss of older needles. Similar effects have been observed in Austria, Switzerland, and France.

Coastal forests of older spruce have also declined visibly since about 1983. (Radial growth reductions began in the mid-1960s.[40]) Damage to root systems has also been observed.

Damage to silver fir has also been documented in West Germany, Austria, Switzerland, and France. Declines in fir due to insects, diseases, and other natural causes have long been reported. Since the early 1970s, symptoms have included yellowing of older needles, necrosis of all needle-year classes, crown thinning, "stork nest" development among younger fir, and secondary branches from "adventitious" buds.[41] "Stork nest" development is normal among white fir trees more than 100 years old: as growth in height is reduced, the tree top becomes flatter, resembling a stork's nest. But now fir trees only 25 to 30 years old follow the same pattern. Adventitious buds appear normally in small numbers on Norway spruce and white fir on the upper side of young branches. But such shoots are much more common in declining trees.[42]

Beginning in the 1980s, signs of injury have also appeared in deciduous trees. On beech, Europe's most common deciduous tree, crown thinning, branch dieback, abnormal leaf morphology, premature foliage discoloration and loss, bark necrosis, growth reductions, and root damage are occurring, especially at high elevations. West Germany's Harz Mountains no longer contain any healthy beech, and even young beech show root and leaf damage—an impediment to regeneration. In some regions, the leaves of declining beech have been characterized as nutrient deficient.[43]

Red Spruce and Fraser Fir in the Eastern United States. Red spruce can be found from North Carolina through Maine. In the Southeast, this tree grows only at higher elevations (1500 to 2000 meters); in the Northeast, its natural habitat ranges from sea level up to 1500 meters.[44]

All along the eastern mountain chain, the radial growth rate of red spruce has declined sharply. In the Northeast, these reductions began in the early 1960s.[45] Continuing into the mid-1980s, they have affected spruce stands of widely differing ages and disturbance histories, and so are unlikely to represent the natural decrease in growth expected of an even-aged stand of second-growth spruce.[46] Indeed, reductions in tree-ring growth may reflect a relatively recent change in the relationship between red spruce trees and their environment.[47]

Many declines of red spruce in the Northeast have been documented over the past century. Major episodes occurred in New York and New England in the 1870s and 1880s.[48] In the Northeast, the most recent decline has been under way for 25 years at elevations over 800 meters, and the number of trees has declined by 50 percent or more.[49] On some mountains, live basal area (the total horizontal area of the tree trunks, measured at breast height) has dropped by 60 to 70 percent. Symptoms observed among high-elevation northeastern spruce include dieback of branch tips from the top and inward from the terminal shoots[50] and a yellowing of the upper surfaces of older needles.[51]

In general, the percentages of dead and severely damaged red spruce increase with elevation.

The overall deterioration of high-elevation spruce in the Northeast has been dramatic over the past several years. Significant reductions in the radial growth of trees has been accompanied by increased mortality. Extensive data gathered at Whiteface Mountain (NY), Mt. Mansfield (VT), Mt. Washington (NH), and other northeast mountains show that above 900 meters, most spruce trees that had lost more than half of their upper-crown foliage in 1982 had died by 1987.[52] Fully 20 percent of the spruce alive in 1982 had died by 1987—a mortality rate about eight times that of spruce at lower elevations. In general, the percentages of dead and severely damaged red spruce increase with elevation. On three New England mountains, almost 60 percent of these trees were dead or dying at 1000 meters, compared with only about 20 percent at 700 meters. Above 900 meters, the red spruce on three mountains in New York, Vermont, and New Hampshire will by 1992 have decreased in abundance by 50 to 80 percent since 1962.[53]

Less severe damage to spruce is also occurring at lower elevations. Foliar symptoms, growth reductions, and some red spruce mortality have been documented throughout the eastern spruce-fir range.[54] On some islands off Maine, spruce exhibit chlorosis, needle loss, "stork's nest" crowns, and extensive crown thinning.[55]

Major damage to conifers has also been observed in the southeastern Appalachian Mountains. Surveys begun during 1983 on Mt. Mitchell in North Carolina and five other southern Appalachian peaks show that growth of red spruce and, to a lesser extent, Fraser fir fell markedly at elevations over 1920 meters beginning in the early 1960s.[56] Damage and ring-width suppression is worst on west-facing slopes. By 1987, almost half of the red spruce and Fraser fir on Mt. Mitchell's west-facing slopes were dead.[57] No significant pathogens have infested the crowns, trunks, or roots of red spruce trees in the southern Appalachians, though fir has historically fallen prey to the balsam wooly adelgid.[58]

High mortality has also been observed in the spruce-fir forests of Tennessee, Virginia, and North Carolina. On a fourth of the 66 thousand acres of spruce-fir forests in these three states, more than 70 percent of the standing trees are dead, much of it Fraser fir killed by the balsam wooly adelgid.[59] In general, crown decline in these high-elevation spruce-fir forests has been in what researchers call class one (0–10 percent needle loss) or class two (10–50 percent needle loss).[60]

In the southern Appalachians and at lower elevations in Maine, needle loss occurs quite differently than at high elevations. Discoloration and loss occur from the inside outward and from bottom to top—much as they do in West Germany.[61]

Yellow Pines in the Southeast. Yellow pines, including loblolly and slash pine, grow over much of the South. Some 42 million hectares (160,000 square miles) of these trees grow in Florida, Georgia, South and North Carolina, and Virginia. Data from the most recent Forest Inventory Analysis (FIA) completed by the Forest Service in the Southeast revealed that the average annual radial growth rates of most yellow pines under 16 inches in diameter have declined by 30 to 50 percent over the past 30 years.[62] Also, pine mortality has increased sharply, from 9 percent in 1975 to 15 percent in 1985.[63] Overall, annual mortality of pine-growing stock has increased by 77 percent. Occurring without any other visible symptoms, abnormal reductions in growth, primarily in natural stands, were termed "worrisome" by the Forest Service.[64]

Although changes in the pine forests can be detected using FIA data, this information is of limited help in determining what causes growth reductions.[65] Most likely, some combination of atmospheric deposition, increased stand density and age, competition with hardwoods, drought, reductions in the water table, changes in soil conditions, and disease is at cause.[66]

Sugar Maple in the Northeast. Sugar maple, a major hardwood species, grows on upland sites in the Great Lake States, New England, Canada's maritime provinces, and Ontario and Quebec. In the late 1970s, sugar maple in southeastern Canada began experiencing crown dieback and elevated mortality rates.[67] Leaves yellowed, autumn colors and leaf drop came prematurely, branchlets began dying from the top of the crown downward, bark peeled on main branches, and trees eventually died. Crown dieback symptoms have also been reported for sugar maple in Vermont, Pennsylvania, New York, and Massachusetts. (Slumps in maple syrup production in 1986 and 1987 are, however, generally attributed to unfavorable weather.) In some regions, yellow birch, American beech, white ash, white spruce, and balsam fir are also showing symptoms of decline.

In Quebec, an aerial survey in 1985 showed that 52 percent of all maple stands in the province showed symptoms of decline, compared with 28 percent the year before.

Sugar maples may be hardest hit in Quebec, where an aerial survey in 1985 showed that 52 percent of all maple stands in the province showed symptoms of decline, compared with 28 percent the year before. Earlier maple declines[68] (in 1934, 1944, 1952, and 1956–58) were brief and ended in recovery. In some parts of Quebec, decline is occurring where no insect attacks have been recorded and no recovery is in sight.[69]

The shallow-rooted sugar maple grows best on deep, fertile, moist, well-drained soils,[70] and it can tolerate only a narrow range of moisture conditions and very little site disturbance.[71] Damage is most severe in highly humid environments or on mountaintops where soils are thin. Affected soils are rather acidic, with low levels of calcium and magnesium and high levels of iron and aluminum.[72] Maple dieback is less pronounced on deep soils without serious nutrient deficiencies.

Crop Damage

The value of U.S. crops, grown on approximately 331 million acres in 1985, totaled $76 billion.[73] In terms of both value and acreage harvested, corn, soybeans, wheat, and hay are the four most important American crops.[74]

As with forests, the health and productivity of these crops are subject to a wide variety of natural and man-made factors, including insects, disease, and air pollution. Heavy use of pesticides notwithstanding, weeds, diseases, and pests still destroy an estimated 37 percent of preharvest crops.[75] Drought, excessive moisture, and frost can reduce a farm's maximum yield by as much as 69 percent in a given year.[76]

Air pollution's effects on crops were first noticed late in the 19th century near such major "point sources" as smelters that emitted large amounts of sulfur dioxide. Only in the last thirty years has more widely spread damage been associated with lower levels of air pollution, primarily ozone.

Natural factors—pests, disease, drought—usually induce visible symptoms that are generally well-understood. Ozone's effects, however, can be visible or invisible, and visible injury may be either acute or chronic.[77] Acute symptoms usually appear within one or two days after high exposures of several hours. Chronic symptoms result from longer-term low-level exposures. Visible effects on leaves include changes in shape, discoloration, and necrosis. Typical ozone injury appears as flecks (small, bleached necrotic areas) or stipples (small pigmented areas).[78] More subtle effects include growth reductions or such physiological changes as those in chlorophyll content.

How much the yields of four important U.S. crops— soybeans, corn, wheat, and peanuts—would increase if current ozone levels were reduced has been estimated by agricultural expert Walter W. Heck and his colleagues.[79] Using validated dose-response models for each crop, researchers estimated that bringing ozone levels down to 0.025 parts per million (ppm) would boost wheat production by 8 percent, soybean production by 17 percent, corn production by 3 percent, and peanut production by 30 percent. These four crops account for about 64 percent of the total cash value of U.S. agricultural output, and the economic benefit from increased production was estimated at about 9 percent of the value of the crops—$3.1 billion in 1978 dollars; $5.4 billion in 1987 dollars.

Bringing ozone levels down to 0.025 parts per million would boost wheat production by 8 percent, soybean production by 17 percent, corn production by 3 percent, and peanut production by 30 percent.

The forest and crop damages summarized here affect a major portion of the country. *(See Figure 1.)* (Only those states with annual agricultural losses of more than $100 million [1987 dollars] are shown.) As the figure makes clear, much of the nation is already experiencing vegetative injury of one form or another. Affected forests are primarily on the west and east coasts while crop injury is occurring through much of the Midwest. In short, the problem is national in scope and will require a national reponse if it is to be solved.

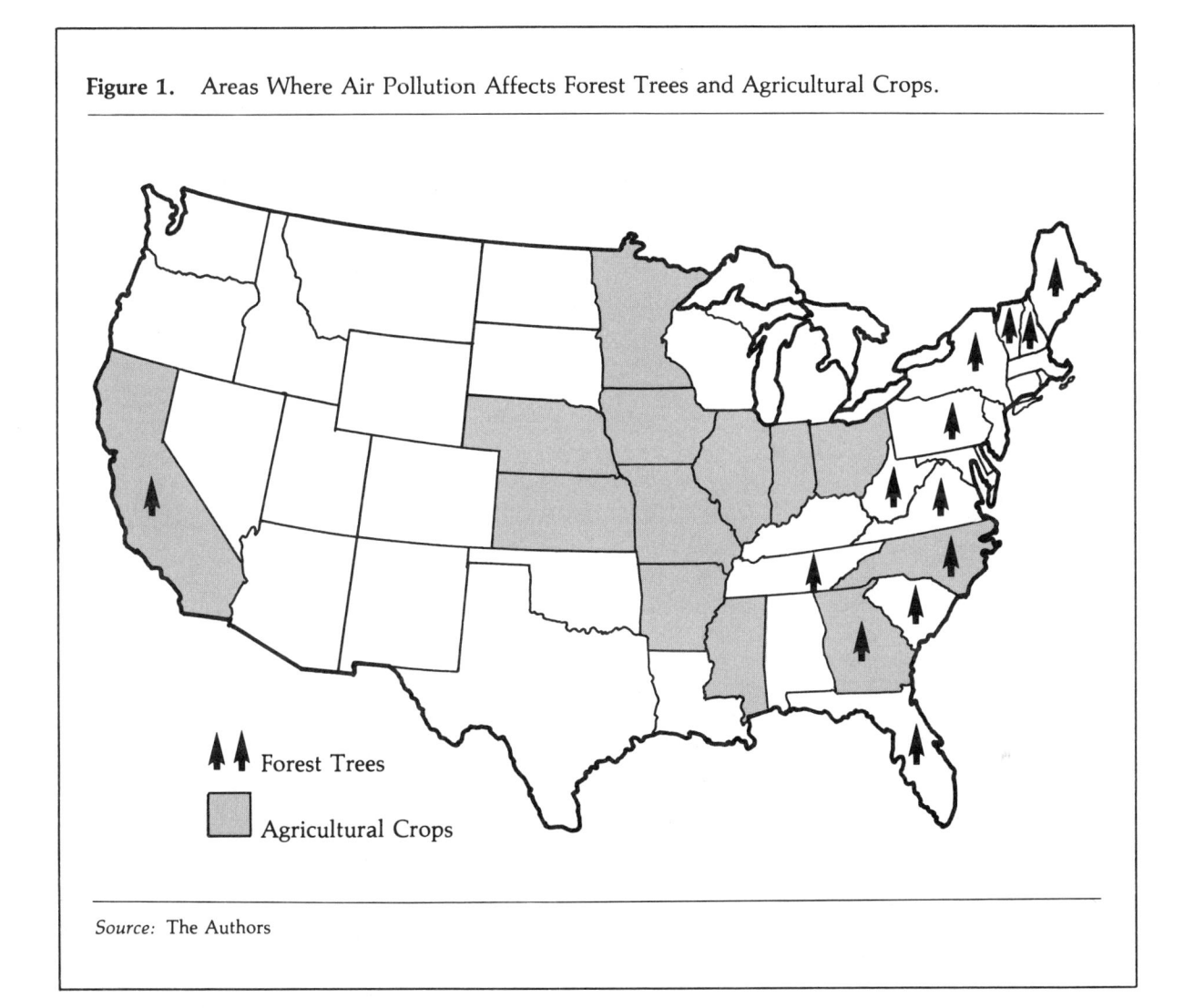

Figure 1. Areas Where Air Pollution Affects Forest Trees and Agricultural Crops.

Source: The Authors

III. Links Between Air Pollution and Damage to Forests

Substantial evidence now indicates that U.S. crops and forests are being damaged. *(See Section II.)* But is there evidence linking air pollutants to these symptoms? In the case of injury to crops and the declines of several species of pines, the answer is a firm yes.

Determining the causes of vegetation injury is difficult because so many factors can influence forest and crop health. Moreover, performing experiments under the field conditions where trees are declining can be arduous. Indeed, it may be impossible to develop a single explanation of the tree mortality plaguing Europe and the United States. If there is any consensus on the causes of the forest declines, it is that multiple stresses—some natural and some man-made—acting together cause observed damages.

Today in Europe, air pollution is believed to *contribute* significantly to observed damage: evidence includes the unique nature of the symptoms, the location of the affected forests in highly polluted areas, and the simultaneous appearance of symptoms in many species and geographical regions.[80] *(See Box.)* In the United States, some declines—for example, those of the Jeffrey, ponderosa, and eastern white pines—have definitely been causally linked to air pollution (ozone). But there is less certainty among researchers about how importantly air pollution figures in other U.S. declines—such as those of the red spruce,[81] Fraser fir, yellow pines, beech, or sugar maple.

Air pollution can injure trees either acutely, as it historically has around smelters and other large point sources, or more gradually at lower concentrations. With the construction of tall smoke stacks in recent decades to reduce local peak concentrations and with continued growth in vehicle use and urbanization, the pollution problem has changed primarily to one of chronic, lower-level exposures that can occur tens or even hundreds of miles from pollutant sources.[82]

The pollution problem has changed primarily to one of chronic, lower-level exposures that can occur tens or even hundreds of miles from pollutant sources.

Today's understanding of how air pollution can damage trees has evolved over the past forty years through research prompted by observed damage at specific sites. Applying this research to recent declines, scientists have identified *many* mechanisms by which air pollution can contribute to tree injury and death.

Air Pollution's Impacts on California's Forests

In the early 1950s, when needle damage to ponderosa pine was first observed in southern California's San Bernardino mountains, healthy and yellowed diseased trees grew side by side.[83] Drought was ruled out because the damage progressed from the bottom to the top and from the inside out (the opposite of drought's normal effects) and trees more drought-sensitive than ponderosa pine showed no symptoms.

Ozone from the nearby Los Angeles basin—known to damage plants—became the primary suspect in the pine decline. To test their suspicion, scientists enclosed various branches of affected trees in chambers and treated them, respectively, with ambient air, carbon-filtered air, and filtered air with measured amounts of ozone added. Branches in the filtered air improved, while the other two showed new or increased injury. These and later fumigation experiments confirmed that ozone caused the damage.

Other studies have since demonstrated that the needles of ponderosa and Jeffrey pine get injured when the 24-hour ozone level is 0.05 to 0.06 parts per million (ppm) from May through September.[84] The mechanism is also well understood: ozone-induced foliar injury, premature needle senescence, and needle loss impairs photosynthesis, reduces carbohydrate production, and decreases radial growth.[85]

Ozone-injured trees are also more susceptible than normal trees to insect attack and root rot.

By examining ozone's effects at increasing distances from the source, scientists have also gauged the relative impacts of decreasing ozone levels. Where levels were high, the trunk growth of ponderosa pine was reduced by as much as 50 percent; growth reductions have also been observed for white fir and California black oak.[86] Jeffrey pine even farther from the source lost needles and died at abnormally high rates. Ozone-injured trees were also more susceptible than healthy trees to insect attack and root rot.

Later surveys showed that trees with visible injury died at higher rates. In plots dominated by ponderosa pine that showed severe injury in 1973, one in three trees had died by 1982–83; only 7 percent had died in stands where only slight injury had been observed earlier.

Research has also ruled out any significant role for several other pollutants. For example, field and controlled fumigation studies carried out near Sequoia National Park to determine whether sulfur dioxide and sulfate air pollution contribute to observed damage suggest that sulfur dioxide levels would have to rise high above present levels before additional injury to existing ozone damage could occur.[87]

Most likely, ozone and other photochemical oxidants will continue to threaten the forests of the southern Sierra Nevada. According to Paul R. Miller, a leading researcher on the decline of California's pines, polluted air from California's Central Valley will continue to imperil the forest itself, its recreation value, and the wilderness heritage of the region's national parks and forests.[88] At the same time, Miller sees less of a threat from ozone and acid deposition inland:

> ...forested areas must be in a consistent downwind direction from urban centers before visible injury from ozone or significant accretion of acidic deposition can be noted. Pacific coastal climates and topography seem to enhance the potential for pollutant accumulation in summer compared to continental climates.[89]

Air Pollution and Forest Declines in the Eastern United States

Over the past 25 years a large number of trees growing in high-elevation eastern forests have unexplainably died. Though the red spruce are the most severely affected, the Fraser fir are also succumbing in large numbers (with no apparent insect-related cause). In the words of the U.S. Forest Service, "something is occurring in

Learning from Europe's Experience

With so many natural and anthropogenic factors affecting forest ecosystems, determining precisely which agents are responsible for killing excessive numbers of trees can be difficult. Indeed, because of the lack of natural controls—as in central Europe where pollution-free forests no longer exist[90]—it is often difficult, if not impossible, to establish cause-effect relationships between air pollution and observed damage.

Forest decline has been a serious issue in Europe, especially in West Germany, for about a decade. Considerable research has been devoted there to examining possible links between air pollution and tree damage. Research in Europe supports the hypothesis that air pollutants, through mechanisms that vary by site, contribute importantly to the current forest decline.[91] Certainly, no other factor considered so far can account for the large number of species affected, the rapid onset of symptoms, the large geographical area affected, and the wide range of associated soil and climate conditions involved.[92]

Damage to the high-elevation fir-spruce forests in the eastern United States began to appear in the early 1960s. Here too, growing evidence suggests that air pollution is a contributor—hardly surprising since the air pollution levels in West Germany compare with those in the eastern United States and the affected tree species are related. As Table 2 shows, the short-term ozone concentrations tend to be higher in West Germany while annual ozone levels are comparable. The acidity of precipitation (measured by its pH value) is also comparable in the two countries. In the eastern United States, however, the cloudwater deposited on high-elevation forests (pH 3.4 to 3.7) is on average about ten times more acidic than the rain. The lowest pH values of cloudwater are in the 2.2 to 2.8 range, making it comparable to vinegar or lemon juice.[93] The amounts of sulfate and nitrate deposited on declining U.S. high-elevation eastern forests—mostly from clouds—are much larger than those in West Germany. In both countries, the amount of acidity and other pollutants deposited dry is highly uncertain.

Of course, comparing the fir-spruce decline in the United States with that in Europe has its pitfalls. European forests are not identical to those in the United States. In Europe, the two conifers most affected are the silver fir *(Abies alba)* and Norway spruce *(Picea abies)*; in the eastern United States, the hardest hit are red spruce *(Picea rubens)* and Fraser fir *(Abies fraseri)*. The soils on which a species grows can also differ greatly, and (like the high-elevation red spruce in the United States) species can be physically and genetically isolated from the main population and thus more vulnerable to some stresses. Yet, while different species (e.g., Norway and red spruce) shouldn't be expected to react alike to given levels of pollutants, the same species in different locations might. Indeed, Norway spruce, planted as an ornamental tree in North America, is showing nearly identical symptoms of stress on both continents.[94]

These differences notwithstanding, the consensus on the mechanisms behind the European decline—direct foliar damage and nutrient imbalance, both induced by air pollution—may well shed light on the declines in the eastern United Sates. These mechanisms provide a consistent explanation of the European problem and account for both wide-ranging symptoms and environmental site-specific observations.

Table 2. Comparison of Air Pollution Levels and Deposition Rates between Forested Areas in West Germany and the United States

Pollutant	Averaging Period	Germany[a]	United States[b]
Ozone (ppb)	Annual Average	36–48	40–50
	Maximum Peak	115–180	85–120
Precipitation (pH)	Annual Av.	4.2–4.6	4.2–4.7
Cloud Events (pH)	Minimum Observed pH	—	2.2–2.8
Total Sulfate Dep. (Wet and Dry) (kg/ha-Yr)	Annual	70–125	100–400
Nitrate Dep. (Wet only) (kg/ha-Yr)	Annual	22–48	50–200

a. Forested areas showing decline symptoms in southern Germany.
b. Various mountains in the eastern United States where tree damage is being observed.

Sources: From Various Tables, "Acid Deposition Studies," Ebasco Services Incorporated, prepared for the Business Roundtable Environment Task Force, (New York: Ebasco, 1986); "Exposure of Forests to Gaseous Air Pollutants and Clouds," Mountain Cloud Chemistry Project, Volker A. Mohnen, June 1987; "Technical Report, Spruce-Fir Research Cooperative," March 1988.

the high-elevation forests that we cannot explain based on our current knowledge of forest and ecosystems processes." More recently sugar maple in New England have also begun to decline, and large numbers of commercial yellow pines in the Southeast have exhibited marked reductions in their annual growth rates. Of concern is the possibility that these various declines are part of a continuing pattern of injury resulting from air pollution.

In the high-elevation Appalachian mountain environments where red spruce and Fraser fir are declining, air pollution levels are high, substantially greater than at nearby lower elevations. At these elevations, the 24-hour average ozone concentrations are typically twice those at lower neighboring elevations.[95] In 1986, about 50 percent of the hourly ozone levels on Mt. Mitchell exceeded 60 parts per billion (ppb), compared to about 20 ppb in nearby Fairview, at a much lower elevation.[96] Average ozone levels are higher on Mt. Mitchell primarily because ozone concentrations do not fall at night as they do at lower elevations.

Hydrogen peroxide—like ozone, a powerful oxidant—also reaches high concentrations in the clouds enshrouding these mountain tops.[97] Acid deposition from rain, clouds, fog, and dry deposition is high. Typical annual sulfate deposition rates at lower elevations in the eastern United States are 20 to 30 kilograms per hectare.[98] In 1986, sulfate deposition on Whitetop Mountain, Virginia, for April through December was 200 to 400 kilograms per hectare, about ten times the annual rate at lower elevations.[99] Similarly, annual hydrogen ion (H^+) deposition—a direct measure of acidity—at low elevations in the eastern United States ranges between 0.3 to 0.6 kilograms per hectare.[100] On Mt. Mitchell, the annual H⁺ deposition rate is between 2.0 and 4.6 kilograms per hectare, roughly ten times that at lower elevations.[101] At high elevations, where the damage is worst, trees may be covered in highly acidic, high-ozone fogs and clouds for up to 3000 hours per year.

On mountains in the Southeast, clouds are the dominant means of wet acid deposition.[102] Cloud samples taken on Mt. Mitchell during 1986 showed a pH varying from 5.4 to as low as 2.2 with a mean pH of 3.4.[103] Minimum pH values for two other eastern mountains were similar: Whiteface Mountain, pH 2.8; Whitetop Mountain, pH 2.6.[104]

Effects of Ozone on Trees. Concentrations of ozone, a well-known phytotoxic agent, are especially high in regions experiencing forest damage.[105] In southern California and in the eastern United States, high concentrations of ozone are the primary cause of the decline in pines. Ozone has also been proposed, along with acid fogs, as a factor in the decline of trees in Europe. Abnormally high nutrient leaching from needles has frequently been observed in young trees fumigated with ozone and acid.[106]

Fogs and mists like those found in high-elevation sites in the eastern United States make it easy for ozone to enter the stomata of leaves and needles. By damaging wall membranes of mesophyll cells containing chlorophyll, thereby reducing photosynthesis,[107] ozone reduces photosynthesis in red oak, loblolly pine, slash pine, ponderosa pine, black oak, sugar maple, and eastern white pine.[108] In studies of eastern white pine trees, researchers found that ozone injured needles and decreased tree growth.[109] For sensitive white pines, chronic ozone exposures reduced annual growth up to 70 percent. The needles of the sensitive trees were 15 to 45 percent shorter than those of more tolerant white pines. Other studies of eastern white pine in Virginia confirmed that ambient ozone levels are suppressing pine growth.[110]

Red spruce, on the other hand, tolerates ozone relatively well.[111] For several hardwood and conifer species, however, the greater the ozone concentration and dose, the greater the reduction in photosynthesis.[112] Indeed, seedling experiments suggest that ambient levels of ozone are stunting the growth of most, maybe all, conifer and hardwood forests in the eastern United States.[113] In seedlings, photosynthetic activity can decline before any observable symptoms appear, so a lack of visible injury does not necessarily indicate a lack of ozone effects.[114]

Besides suppressing photosynthesis, ozone alters the carbohydrate allocation within the tree. Preliminary results from the Boyce Thompson Institute in New York suggest that ambient ozone levels may interfere with the hardening of red spruce, making them more susceptible to winter kill.[115]

Ozone—like drought and temperature extremes—gives nature a push it doesn't need.

Ozone also accelerates the normal leaching of nutrients from foliage. Studies in West Germany of silver fir, Norway spruce, and beech indicate that ozone fumigation damages cell membrane so severely that nutrient loss becomes "uncontrolled."[116] Leaching is especially bad where light intensity is high and nutrients scarce. Nutrient leaching is a normal process and, where soil nutrients are adequate, trees generally replenish them quickly. In this context, ozone—like drought and temperature extremes—gives nature a push it doesn't need.

Arguments against acute ozone exposures as the *primary* cause of the European decline are based on observed damage to the vascular bundles frequently seen in European trees. Such damage is not consistent with acute ozone injury. However, ozone at lower levels, in combination with acid fog, extreme frost, and poor soils does exacerbate tree damage considerably.[117]

Direct Effects of Acid Deposition on Trees. In most experiments on the direct impacts of acid deposition, young conifer and deciduous seedlings have been exposed for periods ranging from weeks to as long as 30 months to simulated acid precipitation of various types and amounts. In one set of experiments, seedlings of tulip poplar, white oak, and Virginia pine growing in poor soils were exposed to simulated acid rain with pH values ranging from 3.5 to 5.6 over 30 months. Researchers found no significant effects in a wide range of health indicators, but cations (such as magnesium, potassium, and calcium) did leach from the upper soils after the most acidic applications (pH of 3.5)[118]

In experiments with red spruce, scientists from Boyce Thompson Institute exposed seedlings separately to mists of nitric acid, sulfuric acid, and a mixture of the two over a pH range of 2.5 to 4.5.[119] Significant foliar damage (20 percent) was observed only from the sulfuric acid mist and only when the pH was at or below 2.6. As the acid droplets dried on needle surfaces, the pH decreased substantially (in one case to a value of 1.6), leading researchers to conclude that acid deposition may be important where evaporation follows wet deposition. In experiments on white ash, seedlings were exposed to pH values ranging from 3.0 to 5.6. The seedlings receiving the most acidic precipitation (pH of 3.0) showed reduced root growth and visible symptoms.[120] In other studies, acid precipitation with pH values ranging from 2.0 to 4.7 was applied to jack pine, and root weight decreased as acidity increased.[121] When exposed to simulated acid rain of pH 2.0, Douglas fir and ponderosa pine seedlings grew limp, almost wilting; needles both old and new became spotted and brown, and many fell off.[122] In a study of acid mist's effects on the leaf cell structure of tulip poplar seedlings, a significant collapse of cells was observed at a pH 2.6. Still other studies showed that at pH levels of 2.6, conifers and hardwoods lost above-ground biomass.[123]

On Mt. Mitchell, experiments with red spruce have demonstrated that acid deposition at ambient levels can damage the stomatal wax plugs of the needles.[124] Since these plugs are believed to minimize water loss and facilitate gas exchange, such injury can retard growth. When groups of five-year old red spruce seedlings were exposed at night to acid mists of varying pH from 2.5 through 5.5, wax-plug damage increased with the acidity of the mist. In similar experiments with Norway spruce, damage (fusion) to wax plugs also resulted.[125] Such injury could, in turn, result in excessive water loss and increased susceptibility to winter stresses.

Other experiments on Mt. Mitchell have shown that rainfall leaches nutrients from needles.[126] In these studies, the chemical composition of rainfall was compared with that of the rain dripping from the needles (throughfall) and of water flowing down the trunks (stemflow). Concentrations of magnesium, calcium, potassium, and sodium in the throughfall and stemflow proved higher than in the rainwater—a finding attributed to leaching from the needles. Elevated levels of sulfate and nitrate were also found in the throughfall and stemflow. Researchers believe that between 40 and 70 percent of the total deposition came from prior cloud- and dry-deposition on the needles.

From these same experiments, a foliar leaching effect could also be inferred. Researchers found that essentially all of the hydrogen ions (H^+) reaching the forest floor came directly from the rainfall. Thus, the acids reaching the needles are being neutralized: the hydrogen from the rain is exchanging with positive cations (magnesium, calcium, potassium, and sodium) in the needles. According to measurements from EPA's Mountain Cloud Chemistry Program, total annual deposition on Mt. Mitchell ranges between 40 and 95 kilograms per hectare of nitrate and 120 to 280 kilograms of sulfate.[127]

Measurements of foliar nutrient levels in mature, healthy-to-the-eye red spruce in the Green and Adirondack mountains, disclosed that magnesium levels were moderately deficient at high elevations.[128] (Potassium foliar levels and soil nutrient levels were not measured.) Studies of foliar calcium, magnesium, and potassium levels at Whiteface mountain showed that declining crown condition correlated with reduced potassium levels in the needles. Potassium levels in the soil appear adequate, so the lower levels of foliar potassium in the declining spruce may be related to increased foliar leaching or to decreased uptake capability. In addition, foliar magnesium levels are also bordering on moderate deficiency.[129]

Ozone and Acid Deposition Acting Together. In still other experiments, conifer needles were exposed to both ozone and simulated acid precipitation. When five-year old Norway spruce trees were fogged twice weekly at a level of pH 3.5 while under continuous exposure to ozone levels of 100 or 300 parts per billion (ppb), the rate of magnesium loss from the needles was 20 percent greater with the 100 ppb ozone exposures than with the controls, and greater still at the 300 ppb level.[130] Similar or greater leaching rates for potassium, calcium, nitrate, and sulfate were observed, leading researchers to conclude that high levels of ozone intensify nutrient loss from needles where acid fog is present.

At precipitation pH levels of 3.0 and above, ozone and acid deposition have few significant direct short-term effects on tree seedlings. Below pH 3.0, however, foliar injury appears and growth decreases.

Together, these studies indicate that at precipitation pH levels of 3.0 and above, ozone and acid deposition have few significant direct *short-term* effects on tree seedlings. Below pH 3.0, however, foliar injury appears and growth decreases, particularly in seedling roots. Notably, during 1986 the pH of cloudwater affecting the above-cloud forests in the eastern United States reached values of 2.2 (Mt. Mitchell), 2.6 (Whitetop Mt.), and 2.8 (Whiteface Mt.).[131]

Acid Deposition's Effects on Soils. Even more important than direct foliar damage are the changes that acid deposition can bring about in soils. Carbonic acid is produced naturally in soils and, by itself, tends to acidify soils. Atmospheric inputs of much stronger acids (primarily sulfuric and nitric acid) add to this continual natural acidification that normally occurs.[132]

How acid precipitation can affect soils and, consequently, tree health has been described by Bernhard Ulrich of the University of Goettingen.[133] In West Germany's Solling forest region, Ulrich estimates, dry deposition accounts for about 75 percent of the total hydrogen ion deposition on spruce stands, and two thirds of the dry deposition comes from the adsorption of sulfur dioxide on needle surfaces.[134]

In some soils, acid precipitation can deplete nutrients by leaching calcium, magnesium, and potassium. The replacement of these important cations by hydrogen ions (H^+) and the mobilization of aluminum can also accelerate soil acidification. Where the input of hydrogen ions is too great to be absorbed by the natural weathering of minerals in the soils that replenishes important nutrients, acidification would be expected. Also, unless enough elements are released in weathering to replace those lost by leaching, nutrient imbalances will eventually occur.

In soils treated with strong acids, the nutrient leaching and aluminum mobilization accelerate while litter decomposition slows (which slows nutrient recycling).[135] "Exchangeable aluminum" (aluminum ions, Al^{3+}, that are not bound within rocks) can damage the fine roots of trees in soils where the ratios of aluminum to calcium or aluminum to magnesium are high.[136] Elevated aluminum levels can block the root uptake of calcium and magnesium, leading to nutrient deficiencies.[137] Excess aluminum can also impair water transport within the tree, increasing sensitivity to drought.

The damage that mobile aluminum can cause to red spruce has been clearly demonstrated in experiments conducted by G.A. Schier, a United States Forest Service researcher.[138] In these tests, seedlings of red spruce and balsam fir were grown in hydroponic solutions containing optimal supplies of nutrients. Aluminum was then added to the various containers in amounts of 25, 50, 100, and 200 mg per liter—much higher concentrations than those found in soils. Allowed to grow for 32 days, the seedlings were then examined for damage and nutrient content. The tops showed little visible effect, though shoot growth in the fir was reduced somewhat. Toxicity symptoms in the roots of the spruce were quite substantial, beginning at 50 mg per liter. The roots of the seedlings exposed to aluminum were thicker and darker brown than the controls. Root length of the spruce decreased as the aluminum concentration increased. In needles and roots alike, concentrations of several important nutrients in both spruce and fir decreased significantly as the amount of aluminum increased. At aluminum levels of 25 mg per liter, foliar nutrient levels (as a percent of the controls) were 52 percent for potassium, 60 percent for magnesium, and 54 percent for manganese. At an aluminum level of 200 mg per liter, the nutrient levels in the needles were 50 percent for potassium, 47 percent for calcium, 35 percent for magnesium, and 33 percent for manganese. Schier found that spruce were more sensitive than fir to aluminum, the root changes observed in the experiment were characteristic of aluminum toxicity, calcium uptake was significantly reduced by aluminum, and the magnesium uptake was reduced particularly sharply. Under natural conditions, Schier concluded, soils with high levels of aluminum reduce plant rooting depth, increase susceptibility to drought stress, reduce plants' ability to use soil nutrients, and make tree roots more susceptible to soil-borne pathogens.

If acid deposition is leading to nutrient leaching from soils and otherwise upsetting the nutrient balance of trees in North America, then observed damage represents a serious, long-term threat.

As noted, leaching probably plays an important role in Europe's forest decline. If acid deposition is leading to nutrient leaching from soils and otherwise upsetting the nutrient balance of trees in North America, then observed damage represents a serious, long-term threat since reversing the symptoms will be difficult at best.

Researchers at Oak Ridge National Laboratory in 1986 evaluated acidic deposition's potential to alter forest soils and thus harm trees.[139] Field experiments indicated that acid solutions will leach cations from soils, especially those near the surface, and magnesium and calcium have typically been found to be more heavily leached than potassium.[140] Still, the results of many acid-irrigation experiments are indeterminate—perhaps, the authors surmise, because they last a comparatively short time. Acid deposition may temporarily increase growth as nitrogen fertilizes trees, but long-term deposition could result in nutrient leaching and eventual decline.[141] So far, none of the experiments have been carried out long enough to test this hypothesis.[142]

According to the Oak Ridge researchers, the soils most susceptible to substantial nutrient leaching and loss of forest productivity are—in technical terms—those with low exchangeable cation reserves (cation exchange capacity of less than 15 cmol(+)/kg), moderate base saturation (20–60 percent), and moderate pH levels (greater than 4.5). Approximately 41 percent of eastern U.S. forests meet these criteria. Actual nutrient leaching rates, stress the researchers, would depend on acid deposition rates, inputs of calcium, magnesium, potassium, and other base cations in deposition, the weathering rates of soils, the mobility of sulfate and nitrate ions, and various biological factors. As for soils within this highly susceptible category where substantial weathering has already occurred, the Oak Ridge team estimated that 18 percent of eastern U.S. soils fit. Summarizing their findings on nutrient leaching and possible aluminum toxicity from acid deposition, the authors found that

"...there appears to be a significant percentage of forested soils (up to 40 percent) that may be subject to a decrease in nutrient status because of leaching by acidic deposition. A few soils, occupying small areas of the eastern United States (up to 4 percent of forested area), may have characteristics that could lead to aluminum toxicity for some forest species."[143]

Obviously, nutrient loss in the eastern United States is potentially significant even though the true magnitude of the problem remains to be determined.

Actual nutrient deficiency in soils in the southeastern United States has been observed in transplant experiments using soils from Mt. Mitchell and Mt. Gibbes.[144] In a study begun in 1986 by North Carolina State University researchers, soils from 2000-meter Mt. Gibbes that were exposed to relatively higher deposition levels were transposed with soils from a 1740-meter site on Mt. Mitchell. Red spruce seedlings were then planted at both elevations in both the transposed soils and control soils from the same sites. Unfortunately, frost heaves during the 1986–87 winter destroyed the experiment on Mt. Gibbes, but in October 1987 the seedlings growing on Mt. Mitchell were compared: 6 percent of those growing on the native, lower soils had died, compared with 93 percent planted in the soil from the higher elevation. The average needle necrosis was 12 percent for those in the lower soil and 92 percent for those in the higher-elevation soil. The average root weight for the seedlings planted in the higher soil was less than half that of those in the lower soil. Clearly, the soils from the higher elevation could not sustain the spruce seedlings.

Needle analysis disclosed that the seedlings growing on the high-elevation soils had much lower nutrient levels than those growing on the low-elevation soils. The average uptake of potassium, magnesium, and calcium by the deficient trees was only 35, 47, and 49 percent, respectively, of that of the healthy seedlings. Soil analysis revealed that the levels of these nutrients were much lower in the high-elevation soils than in those from lower elevations: the base saturation (a measure of nutrient availability) was only 11 percent in the high-elevation soils, compared to 22 percent in the lower soils. (By this measure, the higher the value, the more nutrients are available.) The pH and aluminum levels in the soils were approximately equal. Both "needle mass per tree" and "nutrient uptake per tree" turned out to be directly proportional to the base saturation of the soil: the more nutrients available, the more the trees absorbed them and grew. Also, as the ratio of aluminum to calcium rose, the needle mass of the trees dropped sharply, suggesting the aluminum-calcium antagonism at work.

Evidence also suggests that soil acidification is occurring in the northeastern United States and that trees there may be experiencing nutrient deficiency. Measurements of soil pH made throughout New York's Adirondack Mountains in the early 1930s were repeated in 1984 by S.B. Andersen of the University of Pennsylvania.[145] Generally, soils with a pH over 4.0 in the 1930s had grown more acidic over the years and now have pH levels between one-half and one unit lower.

Also in the Adirondacks, soil changes in the Huntington Forest, where spruce die-back is occurring, have been studied. In a thoroughgoing data review, George Tomlinson, a tree chemist with Domtar, Inc., Research Centre in Quebec found very large net calcium losses from the soil as a result of acid leaching.[146] The ratio of aluminum to calcium in the near-surface soils was so high that trees' fine roots were being damaged. In related research, Walter Shortle and Kevin Smith of the U.S. Forest Service examined aluminum's role in blocking calcium uptake by red spruce in the Northeast.[147] These researchers concluded that aluminum in the soils limits the calcium supply of spruce trees, thereby contributing to their decline, and that continued acid deposition to these soils will make matters worse.

Excess Nitrogen Deposition. Nitrogen, absolutely vital to tree growth, is frequently the limiting nutrient in U.S. forests. Yet, some evidence indicates that nitrogen deposition is excessive on the mountains where red spruce are declining. By one estimate, total wet deposition of nitrogen at a high-elevation New Hampshire site is seven times as great as at low elevations.[148]

Excessive nitrogen deposition can harm forests.[149] Vigorous growth stimulated by nitrogen fertilization may lead to nutrient deficiency if other nutrients are not sufficient. Nitrogen compounds can adversely alter physiological and anatomical development (e.g., winter hardening). Excessive quantities are believed to increase trees' susceptibility to freezing or desiccation in winter—a suspected contributor to red spruce decline.[150] Cold temperatures, whether from an early frost before winter hardening is completed, an exceptionally cold spell in winter, or a cold period following an early thaw, may harm red spruce.[151]

Needle browning and loss observed on red spruce in Vermont and New York have been attributed to winter injury.[152] Researchers have found that nitrogen levels in the damaged foliage of red spruce and balsam fir are significantly greater at higher elevations than at lower elevations, as are the ratios of nitrogen to calcium, magnesium, and manganese. A related finding was that calcium, magnesium, and phosphorous concentrations in northern New York and Vermont appear to be lower than values reported elsewhere for red spruce.[153] (Magnesium levels bordered on deficiency.) In other experiments involving young Norway spruce heavily fertilized with nitrogen, marked magnesium deficiency and needle-tip yellowing was observed.[154]

Ample evidence shows that the various damage mechanisms at work in the United States are also at work in European forest declines.

Ample evidence shows that the various damage mechanisms at work in the United Sates are also at work in European forest declines. In a recent review of forest decline, Bernhard Prinz of the University of Essen, summarized West Germany's current understanding of *neuartige Waldschaeden* in Europe as follows:

• *The cause of the short-term appearance (and recovery) of damage involves climate as a triggering or synchronizing factor.*

• *The causes for the long-term temporal development of damage involve the upward trend of ozone concentration during the last few decades and the continuous loss of nutrients in soil by acid deposition.*

• *The causes for the spatial distribution of damage involve the increase of ozone concentration as a daily average with increasing altitude and the natural differences in the nutrient supply from soils.*[155]

If Prinz is right, long-term exposure to air pollution leads to both direct tree damage and shortages of nutrients vital to tree growth. In particular, ozone and acid deposition may act together to leach magnesium, potassium, and calcium from foliage. Some of these nutrients will also be leached from the rooting zone along with excess nitrate and sulfate that has not been absorbed by the vegetation. Where soils are shallow, these nutrients may not be able to be recycled or replaced fast enough from the soils or decaying surface litter, so trees will weaken and die from secondary stresses. Where the soils are richer, acid deposition can over time alter the ratios of nutrients in the soil so drastically that trees weaken or die because they cannot absorb all the nutrients they need.

Reinhard Huettl of the University of Freiburg in West Germany has described the European understanding of *Waldschaeden* in more detail.[156] The chain of events begins with air pollution, primarily acid deposition and ozone. These pollutants leach such important nutrients as magnesium, calcium, and potassium from the needles and leaves in the forest canopy. *(See Box, page 24.)* Ozone weakens the cell membrane system and, in combination with acid deposition, exacerbates the leaching of important nutrients.[157] To compensate for these losses, trees try to take up more of the depleted nutrients from the soil. But even as foliar leaching progresses, both natural soil acidification and that from acid deposition is taking place. The hydrogen ions (H^+) from the acid precipitation that falls directly to the ground are exchanged with the nutrients in the soils (magnesium, calcium, potassium, etc.). These nutrients are then leached from the soils with the excess nitrate and sulfate not absorbed by the trees. Acidification accelerates as the trees attempt to replace the nutrients leached from the canopy with those from the soils and in so doing they release additional hydrogen ions (H^+) to the soil (which maintains a neutral charge balance). In highly acidified soils, aluminum can be mobilized and the uptake of magnesium and calcium can be blocked where aluminum-to-magnesium or aluminum-to-calcium ratios become sufficiently high.[158] On more alkaline soils, acids will be neutralized, but neutralization can increase the calcium-to-potassium ratio, blocking potassium uptake.[159] These damaging processes can be exacerbated where nitrogen deposition is very high: nitrogen fertilizes the trees year round and can stimulate tree growth where magnesium, potassium, or calcium supplies are limited.

According to this viewpoint, air pollution—primarily ozone and acid deposition—leads over time to both direct foliar damage and the leaching of nutrients from both trees and soils. Which nutrients will be deficient depends on soil conditions.[160] As the trees weaken, climatic extremes or other natural stresses can cause further weakening and damage, nutrient deficiency, and tree death.

In Manion's framework *(See Section II)*, soils susceptible to nutrient alteration represent an important predisposing stress upon the trees they support. Air pollution is then considered as an inciting factor. Insects and climate extremes may be contributing factors.

From the accumulated evidence presented above it is clear that—where air pollution is an important factor—the mechanisms of forest decline will vary with location. The symptoms displayed by various tree species depend on both the soils and the levels and nature of the deposited pollutants. In forests with acidic soils, symptoms will evolve differently from those in forests with a high buffering capacity. The common thread among the declines is that the trees will eventually show direct damage or develop a shortage of one or more of the nutrients needed for growth. Understanding exactly what is happening at a specific site requires careful examination of both the nutrient status of the trees and the chemistry of the soils.

The Role of Macro-Nutrients in Tree Decline

Several macro-nutrients are vital to a tree's health and growth, and well-known symptoms arise predictably in their absence. Three key nutrients whose deficiency in soils has been linked with air pollution are magnesium, potassium, and calcium.

Magnesium (Mg). Magnesium is a constituent of chlorophyll, which is needed to convert carbon dioxide into organic matter. Magnesium is mobile in trees. In conifers, magnesium from older needles moves to the newer outer needles if there is a deficiency in the soils. The older needles then turn yellow (a condition called chlorosis) and eventually die. In short, magnesium deficiency results in needle loss from the trunk outward and from the base upwards.

Potassium (K). Potassium is essential to tree growth. Without it, roots could not push their way through the soil; nor could the tree's bark expand as the tree grows radially outward. Potassium, like magnesium, is highly mobile, and without sufficient supply foliage begins to yellow, much as it does when magnesium is deficient.

Calcium (Ca). Calcium is essential to the formation of cell walls and to the tree's radial and vertical growth. Calcium pectate forms the active cell walls of the cortex of the fine roots through which inorganic nutrients and water enter the tree. It is not mobile and moves to new growth only when supplies in the soil are adequate. When there is inadequate calcium, root development is poor, growth is reduced, and foliage is lost from the top down and inward from the ends of the branches—the opposite to the pattern observed in magnesium and potassium deficiencies.

Source: George H. Tomlinson, "Nutrient Deficiencies and Forest Decline," Paper presented at Canadian Pulp and Paper Association Annual Meeting, Montreal, Jan. 29, 1986.

IV. Links Between Air Pollution and Damage to Crops

Historically, sulfur dioxide and hydrogen fluoride were the first air pollutants known to damage vegetation.[161] Typically, they killed most vegetation within a few miles of smelters, electric power plants, or other large point sources. More widespread damage to vegetation from photochemical air pollution was first recognized in 1944 in the Los Angeles area. According to Walter Heck, research leader of the U.S. Department of Agriculture's Air Quality Research Program, photochemical oxidants (primarily ozone) now injure and damage crops across most of the United States.[162] Other air pollutants implicated in crop damage include peroxyacetyl nitrate (PAN), nitrogen dioxide (NO_2), and ethylene.

The air pollutants of greatest national concern to agriculture today are ozone (O_3), sulfur dioxide (SO_2), nitrogen dioxide (NO_2), and sulfates and nitrates. Of these, ozone is by far the most worrisome.

The air pollutants of greatest national concern to agriculture today are ozone (O_3), sulfur dioxide (SO_2), nitrogen dioxide (NO_2), and sulfates and nitrates. Of these, ozone is by far the most worrisome. In studies of how ozone and sulfur dioxide affect soybeans, potatoes, and beans, researchers found no significant effect from SO_2 and no interaction with ozone unless the SO_2 exposures were much greater than those typically found in the United States.[163] In contrast, the possible role of acid deposition at ambient levels remains to be determined. At present levels, most studies indicate, acid deposition does no identifiable harm to foliage.[164] But, at lower-than-ambient pH levels, various impacts include leaf spotting, acceleration of epicuticular wax weathering, and changes in foliar leaching rates. When applied simultaneously with ozone, acid deposition also reduces a plant's dry weight. Unfortunately, the impacts of the dry deposition of acid-forming substances on crops cannot be assessed for several years: scientists just don't have enough information yet.[165]

Identifying Ozone's Impacts on Crops

Walter Heck has developed a three-stage procedure for estimating the total impact of air pollutants on crops.[166] First, the response of various crops to specific pollutant concentrations over time must be empirically determined and analytically described. Second, a crop-inventory data base must be devised. And, third, the crop inventories and findings on responses must be combined with a suitable air pollution data base to estimate total crop losses.

Evidence on how crops respond to pollutants can be gathered using several techniques. For example, crops can be subjected to varying controlled concentrations of pollutants to determine a dose-response relation. Alternatively, protective chemicals could be applied to some plants to assess the pollutant's impact on unprotected plants. If a pollution-resistant cultivar of a crop is available, then the impact of varying pollutant levels can be assessed by comparing its hardiness with that of non-resistant cultivars. Scientists can also use open-top chambers to expose crops to controlled amounts of pollutants. Researchers have also estimated the lowest ozone levels and duration times that can damage leaves or reduce plant growth and yield.[167]

Research has shown that ozone enters crops through the microscopic openings on the leaves and that ozone exposures can lead to either visible or more subtle effects.[168] The plant won't suffer injury if ozone levels are so low that it can detoxify the gas and its metabolites or otherwise repair the damage. Peaks in ozone concentrations are more likely to harm plants. Cell damage that is not repaired or compensated for ultimately leads to visible effects (such as changes in morphology or color or to tissue death) or secondary effects (such as reduced plant growth, decreased yield or crop quality, and alterations in susceptibility to stress.)[169]

Visible symptoms can arise from either acute or chronic exposures. Acute symptoms appear within a day or two of short exposures (measured in hours) to high concentrations. Typical symptoms of acute ozone exposure include chlorosis, flecking, and stippling. Chronic exposures can cause chlorosis or other color changes and, eventually, cell death. But the symptoms of chronic ozone exposure are not reliable in diagnosis since they are easily mistaken for injury from diseases, insects, and other natural stresses.

So far, acute injury is the only certain form of ozone damage, and researchers now know how it affects the growth and yield rates of many plants. Yet, as Figure 2 shows, some low-level ozone concentrations and exposure times (those below the broken line on the graph) appear essentially harmless.[170] Indeed, the lower limit for reducing plant performance is an ozone concentration of 0.05 ppm for several hours daily for more than 16 days. For exposures of 10 days, the ozone threshold increases to about 0.10 ppm and, for six days, to about 0.30 ppm.[171]

Quantifying the Impacts

Exposures to ozone can have a number of measurable effects on plants generally and on crops specifically. Growth and yield fall off when ambient concentrations reach certain levels, and the quality of the usable product may also be impaired. These impacts were studied intensively between 1980 and 1987 as part of the government-sponsored National Crop Loss Assessment Network (NCLAN) review of ozone's effects on corn, cotton, peanuts, sorghum, soybean, wheat, alfalfa, barley, clover, tomato, and tobacco.

The goal of the NCLAN research was to estimate crop yield losses resulting from ambient ozone concentrations above naturally occurring levels. "Yield loss" was defined to include changes in plant appearance; losses in weight, number, or size of the plant parts harvested; and any changes in physical appearance, chemical composition, or ability to withstand storage.[172]

The dose-yield data collected for the eleven crops mentioned above were fitted to an empirical, non-linear model. This so-called Weibull model was then used to predict the crop yield losses as a function of the seasonal seven-hr/day mean O_3 concentration. *(See Table 3.)* As Figure 3 shows, mean seasonal 7-hour maximum ozone concentrations at rural locations vary remarkably little across the country. (In an ozone-free atmosphere, yield losses would be zero.) Figure 4 shows the yield losses for corn, wheat, cotton, soybeans, and peanuts as a function of the seasonal seven-hr/day ozone mean. Figure 5 indicates that 8 of 37

Figure 2. Relationship Between Ozone Concentration, Exposure Rate, and Total Exposure Period for Observed Reductions in Plant Growth.

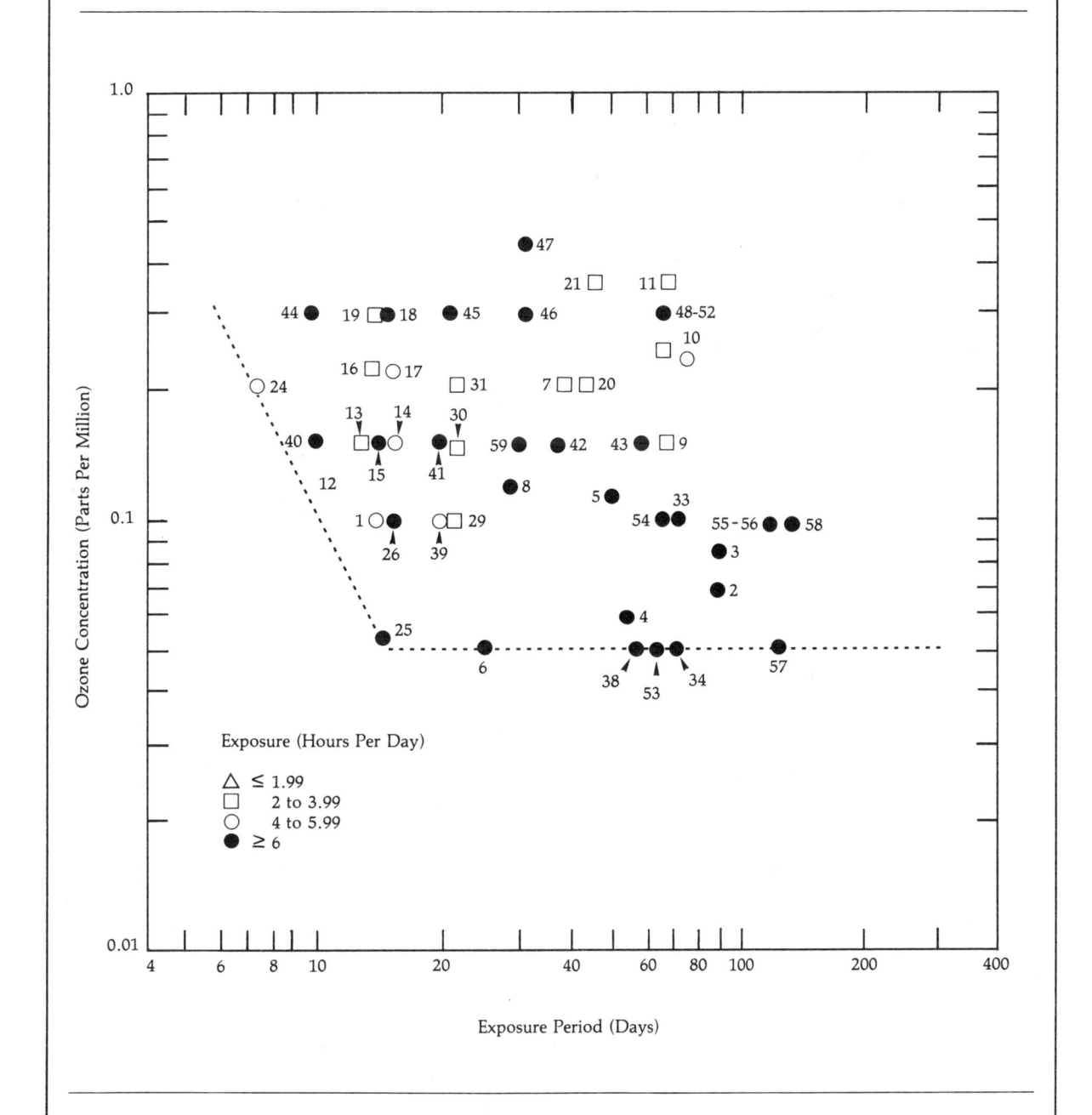

Source: U.S. EPA, "Air Quality Criteria for Ozone and Other Photochemical Oxidants, Vol. III., August 1986, page 6-226.

Figure 3. Average Daily 7-Hour Maximum Ozone Concentration (Parts Per Billion) in Rural Areas During the Growing Season.

Source: Interim Assessment, Vol. IV, National Acid Precipitation Assessment Program, September 1987, page 6-5.

Figure 4. Predicted Yield Losses for Corn, Wheat, Cotton, Soybeans, and Peanuts as a Function of the 7-Hour Seasonal Ozone Concentration.

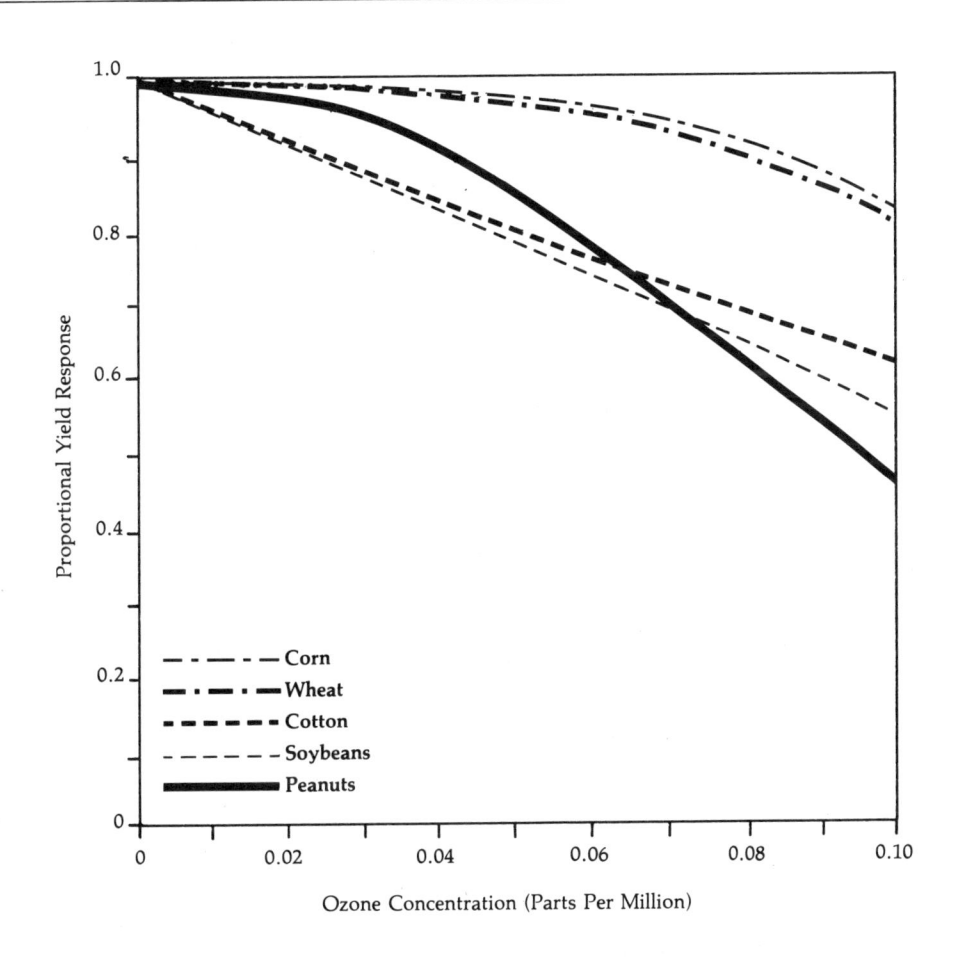

Source: "Assessment of Crop Losses From Air Pollutants in the United States," Walter W. Heck, to be published in World Resources Institute book, 1989.

Figure 5. Number of Crop Species or Cultivars (From a Total of 37) Showing a 10-Percent Yield Loss at Indicated 7-Hour Seasonal Mean Ozone Concentrations.

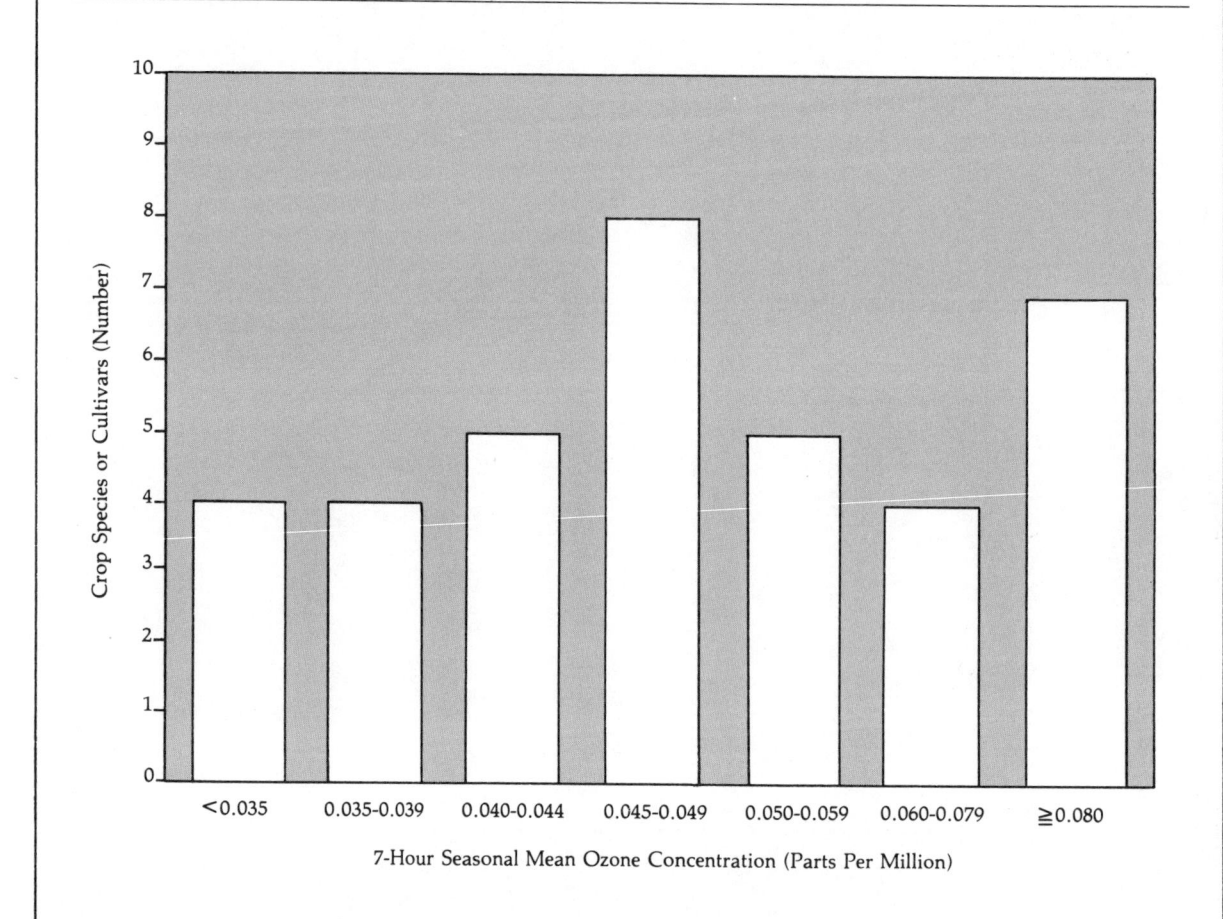

Source: U.S. EPA, "Air Quality Criteria for Ozone and Other Photochemical Oxidants, Vol. III., August 1986, page 6-233.

Table 3. Predicted Yield Losses (Percent) at Several Seasonal 7-Hr/Day Mean O_3 Concentrations

Species	Concentration (ppm)			
	0.04	0.05	0.06	0.09
Barley	0.1	0.2	0.5	2.9
Bean, Kidney	11.0	18.1	24.8	42.6
Corn	0.6	1.5	3.0	12.5
Cotton	4.0	6.9	10.0	20.0
Peanut	6.4	12.3	19.4	44.5
Sorghum	0.8	1.5	2.5	6.5
Soybean	7.3	12.1	17.0	30.7
Tomato	0.7	1.7	3.6	16.0
Winter Wheat	3.5	6.9	11.1	27.4

Source: Walter W. Heck "Assessment of Crop Losses From Air Pollutants in the United States," in *Multiple Air Pollutants and Forest and Crop Damage in the U.S.* (working title) Yale University Press, Spring 1989.

crops or cultivars are reduced by 10 percent at ozone concentrations between 0.045 and 0.049 ppm. More than half the crops are predicted to show a 10-percent loss at seven-hour seasonal mean concentrations below 0.05 ppm, the level prevailing in most agricultural regions.[173] About 11 percent of the species or cultivars show a 10-percent loss below 0.035 ppm.

Estimating Productivity Losses

Data from various research programs clearly show that ambient levels of ozone are high enough in parts of the United States to impair plants' growth and yield.[174] The economic losses from current ozone levels can be estimated by combining crop inventory data, ambient ozone data, and the Weibull model linking various ozone levels to yield losses. An economic model is also required to convert yield losses into economic losses. (Simple economic models estimate economic losses by multiplying crop losses by a fixed price per commodity. More complex models take market behavior more fully into account.)

Ambient ozone levels in most agricultural regions are about 0.05 ppm.[175] In an estimate cited by Walter Heck, yield losses for soybeans, corn, wheat, and peanuts were calculated by comparing present ozone levels to a background level of 0.025 ppm. For these four crops, approximately three billion dollars of productivity would be gained if current maximum ozone concentrations were reduced to 0.025 ppm.[176] Compared with a background ozone level of 0.025 ppm, present ozone levels probably lead to yield losses in U.S. crop production of 5 to 10 percent.[177]

Present ozone levels probably lead to yield losses in U.S. crop production of 5 to 10 percent.

Reviewing economic assessments of crop losses to help develop criteria for an ozone ambient air quality standard, EPA found two studies to be "reasonably comprehensive" in their estimates.[178] In the first, yield losses for corn, soybeans, cotton, wheat, and peanuts were estimated.[179] According to the researchers, reducing ambient ozone levels (seasonal 7-hour average) to 0.04 ppm would result in a $2.1 billion net benefit (1987 dollars). If ozone concentrations increase to 0.08 ppm, net losses would total $5.2 billion (1987 dollars). In the second review, losses for six major crops that account for over 75 percent of U.S. crop acreage—corn, soybeans, wheat, cotton, grain sorghum, and barley—were examined.[180] By reducing 1980 ozone levels by 25 percent, benefits of $2.3 billion (1987 dollars) would be realized.

V. Air Pollution in the U.S.: Sources, Trends, and Proximity to Forest and Agricultural Damages

As documented in Section II of this report, damages are occurring to forests and crops across the country. Indeed, almost half of the lower-48 states are experiencing some such loss, and these injuries have either already been linked to air pollution or are currently under suspicion. *(See Figure 1.)* The origins of these implicated pollutants are reviewed below, in terms of both geography and type of activity. This review makes it clear that the states experiencing tree and crop damages are generally adjacent to, or downwind from, the principal precursor sources for acid deposition or ozone. Combined with the scientific evidence already presented, this geographic proximity adds yet further weight to the proposition that air pollution contributes to damage to forests and crops.

The nature and sources of air pollution in the United States have changed considerably over the past century. With U.S. industrialization and the substitution of coal for wood, wind, and animal power in the latter 19th century, intense local air pollution problems developed, mainly from emissions of particulates and sulfur dioxide. Within the past four decades or so, with rapid growth in transportation and the use of tall stacks to reduce local pollutant levels, air pollution problems have become regional, national, and international in scope. In the post-war era, attention has gradually shifted from the primary pollutants directly emitted into the atmosphere (sulfur dioxide, nitrogen oxides, particulates, carbon monoxide) to secondary pollutants (sulfates, nitrates, ozone). Of particular concern here are emission trends and the conversion of primary pollutants into the secondary pollutants that pose public health and environmental risks.

Trends in Pollutant Emissions

Sulfur dioxide (SO_2), nitrogen oxides (NO_x), and volatile organic compounds (VOCs) are the three classes of air pollutants of primary concern in reducing damages to crops and forests. Total emissions of these pollutants have grown considerably during the twentieth century, reflecting long-term trends in both population growth and the substitution of inanimate energy sources for animal and human labor.

Sulfur Dioxide. Emissions of sulfur dioxide (SO_2) result principally from fossil fuel combustion.[181] Emissions increased from about 9 million metric tons per year (MMT/Y) in 1900 to about 21 MMT/Y in the late 1920s. After declining during the Great Depression, they rose again during World War II. In the 1950s, they declined as transportation and industrial coal markets were lost to oil and natural gas. *(See Figure 6.)* Sulfur dioxide emissions peaked in 1973 at about 30 MMT/Y and had declined by 1986 to about 21 MMT/Y.

Sulfur compounds are released to the atmosphere from both natural and man-made sources. Natural sources—soils, vegetation, oceans, and volcanoes—together contribute less

Figure 6. Historical Trends in Sulfur Dioxide Emissions in the United States by Source Category.

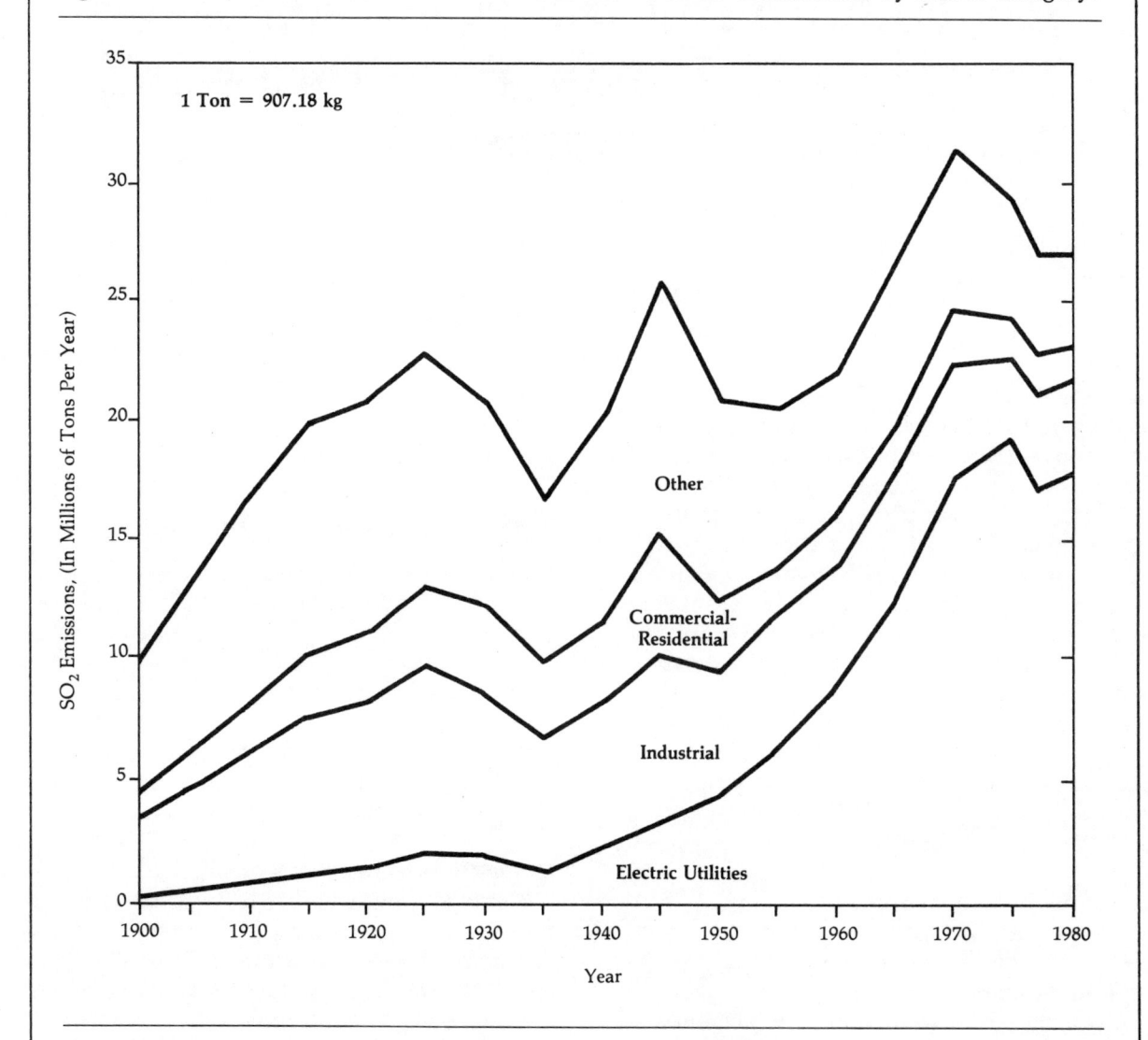

1 Ton = 907.18 kg

Other

Commercial-Residential

Industrial

Electric Utilities

SO_2 Emissions, (In Millions of Tons Per Year)

Year

Source: "Historical Emissions of Sulfur and Nitrogen Oxides in the United States from 1900 to 1980," G. Gschwandtner et al., Journal of the Air Pollution Control Association, Feb. 1986, Vol. 36, No. 2.

than 4 percent to total U.S. sulfur emissions and less than 6 percent to combined U.S. and Canadian emissions.[182]

The principal anthropogenic sources of SO_2 emissions have gradually changed over the century. *(See Figure 7.)* Coal used in heating, industry, and transportation was the primary source of both energy and SO_2 emissions near the turn of the century. Smoke stacks were of only modest height at that time, so SO_2 concentrations were high, but fairly localized.

For reasons of cleanliness and convenience, coal has been displaced over the years from its traditional markets. Diesel locomotives replaced steam engines, and oil and gas furnaces superseded coal boilers in commerce and residences. The largest growing market for coal, however, has been for electric power production. By 1986, some 85 percent of all coal, which accounts for 75 percent of national SO_2 emissions, was burned to generate electricity.[183]

Electric utilities' growing use of coal for fuel and tall stacks to disperse emissions has helped spread the acid deposition problem.

The electric utilities' growing reliance on coal along with the use of tall stacks to disperse emissions have helped spread the acid deposition problem. As Figure 8 shows, more than two thirds of all sulfur dioxide emissions come from stacks over 73 meters (240 feet) high. While the use of tall stacks may lower SO_2 concentrations locally, it can also lead to greater dry and wet deposition of sulfur compounds, including sulfuric acid, far from the power plants.

Most sulfur dioxide is emitted in the eastern half of the country. *(See Figure 9.)* Indeed, nine of the ten states that are largest sources of sulfur dioxide are eastern. These ten states accounted for 59 percent of national emissions in 1985. *(See Table 4.)*

Table 4. Ten Largest Emitters of SO_2 in 1985 (Millions of Metric Tons, and percent of total emissions)

Ohio	2.43	11.3%
Indiana	1.67	7.8
Pennsylvania	1.5	7.0
Illinois	1.27	5.9
Texas	1.01	4.7
Missouri	1.0	4.7
W. Virginia	0.99	4.6
Georgia	0.96	4.5
Kentucky	0.9	4.2
Tennessee	0.89	4.1

Total: 12.62 Million Metric Tons (59% of total emissions)

Source: NAPAP Interim Assessment, Vol. II, pp. 1–35

Nitrogen Oxides. Nitric oxide (NO) is the principal man-made nitrogen pollutant emitted to the atmosphere. This gas, comprising about 93 percent of anthropogenic nitrogen oxide emissions, is produced when nitrogen in the air combines with atmospheric oxygen during fuel combustion. It is then quickly converted to nitrogen dioxide (NO_2). The principal natural sources of nitrogen oxides (which include both nitric oxide and nitrogen dioxide) are lightning and biological processes in soils. Hard to measure, natural processes are believed to contribute about 10 percent of total NO_x releases.

Atmospheric scientists have long recognized that within minutes of its release nitric oxide (NO) reacts with ozone and establishes an equilibrium with ambient nitrogen dioxide (NO_2) and that the ratio of the concentrations of NO_2 to NO determines local ozone levels. Principally because of its important role in ozone formation, NO emissions are of great concern.

Figure 7. Historical Trends in Sulfur Dioxide Emissions in the United States by Fuel Type.

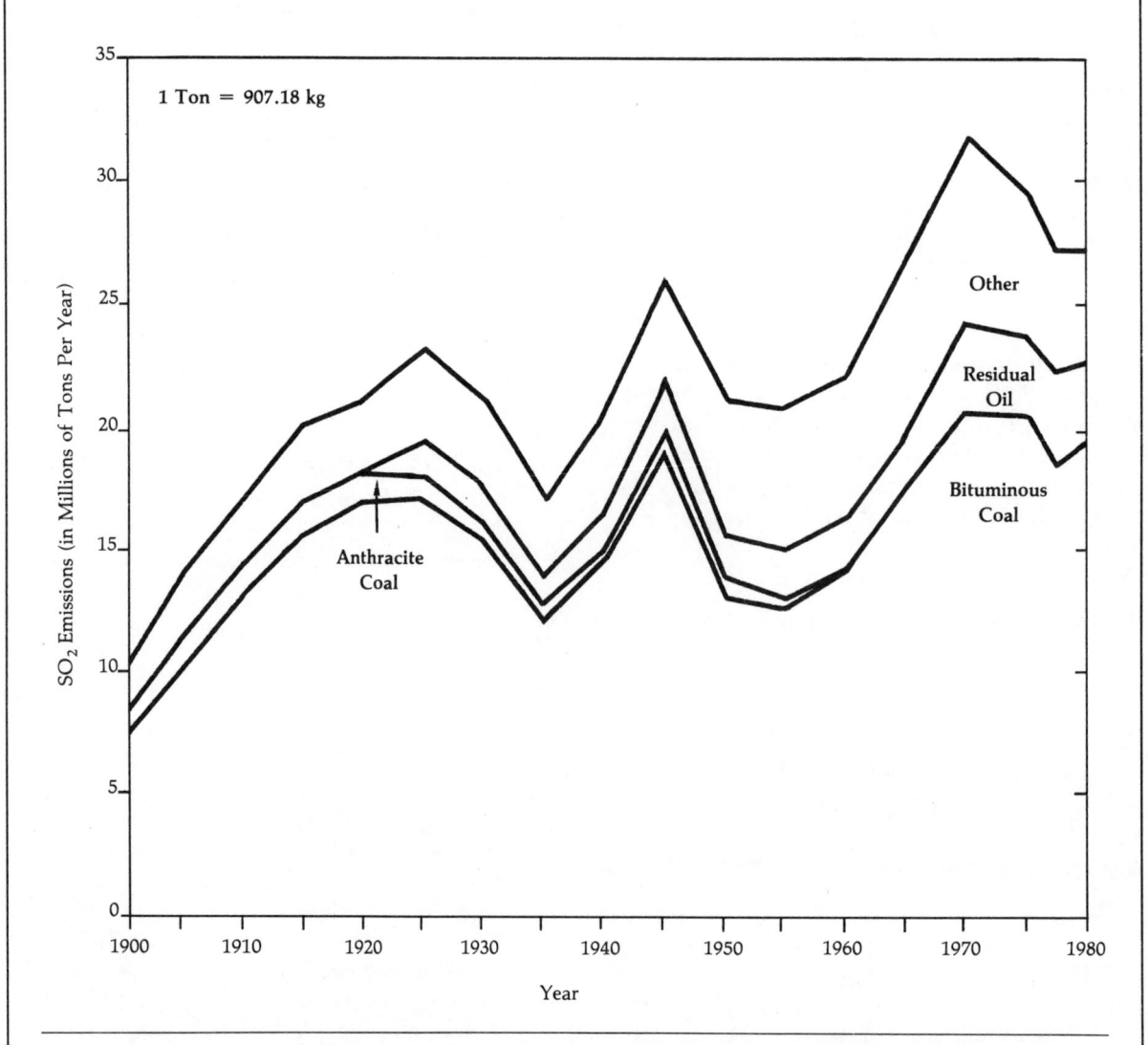

Source: "Historical Emissions of Sulfur and Nitrogen Oxides in the United States from 1900 to 1980," G. Gschwandtner et al., Journal of the Air Pollution Control Association, Feb. 1986, Vol. 36, No. 2.

Figure 8. Historical Trends in Sulfur Dioxide Emissions in the United States by Stack-Height (Meters).

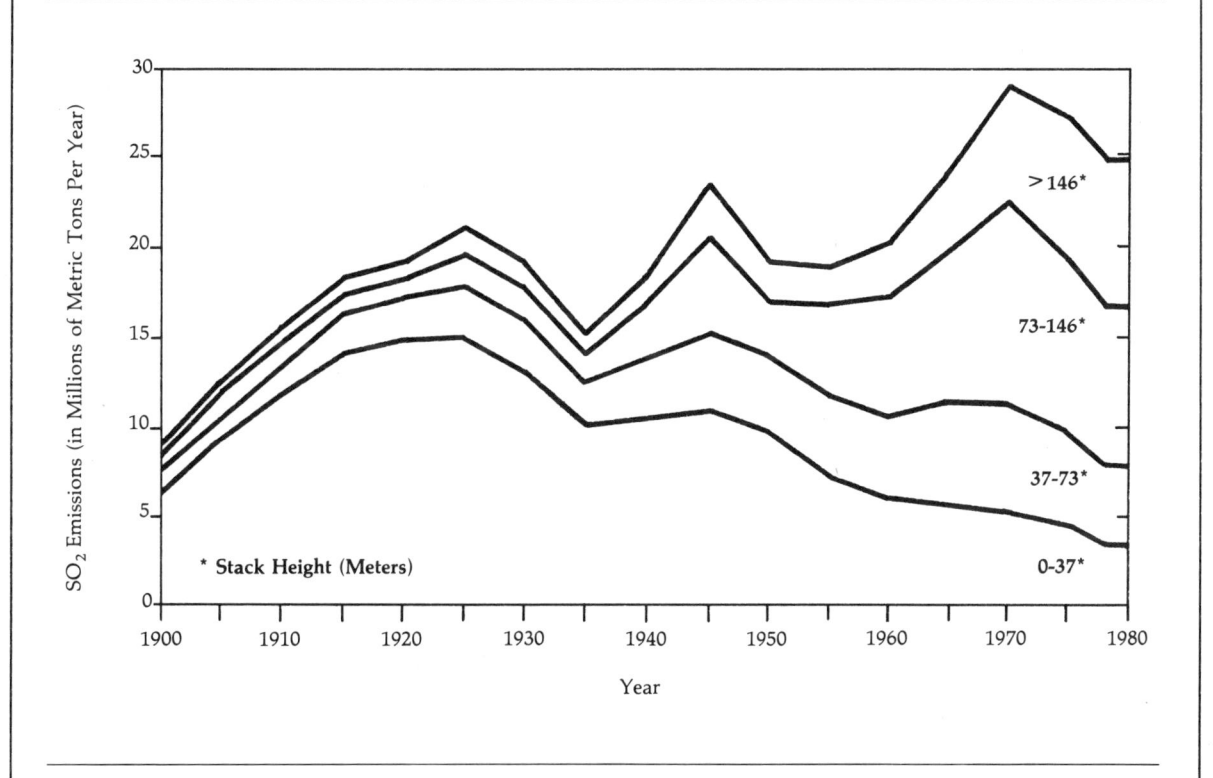

Source: Interim Assessment, Vol. II, National Acid Precipitation Assessment Program, September 1987, page 1-26.

Since nitrogen oxides (NO_x) are formed during combustion, emissions have increased in parallel with fuel burning during this century. *(See Figure 10.)* EPA estimates total NO_x emissions in 1985 at 20 million metric tons per year (MMT/Y).[184] As Table 5 shows, transportation activities account for about 45 percent of all emissions, followed by electric power plants at 28 percent.

Unlike SO_2 releases, which come primarily from coal combustion, NO_x pollution results when any fuel is burned. *(See Figure 10.)* Nitrogen oxides also tend to be generated much closer to ground—from highway vehicles—than

Table 5. Sources of NO_x, 1980

Highway Vehicles	35%
Other Transportation	10%
Power Plants	28%
Industrial combustion	19%
Other Sources	8%
TOTAL	100%

Source: NAPAP Interim Assessment, Vol. II, pp. 1–44

Figure 9. Density of Sulfur Dioxide Emissions from Man-Made Sources in 1980.

Metric Tons (Per Square Kilometer)
- ☐ 0–0.12
- ▨ 0.12–1.50
- ■ >1.50

Source: Interim Assessment, Vol. II, National Acid Precipitation Assessment Program, September 1987, page 1-33.

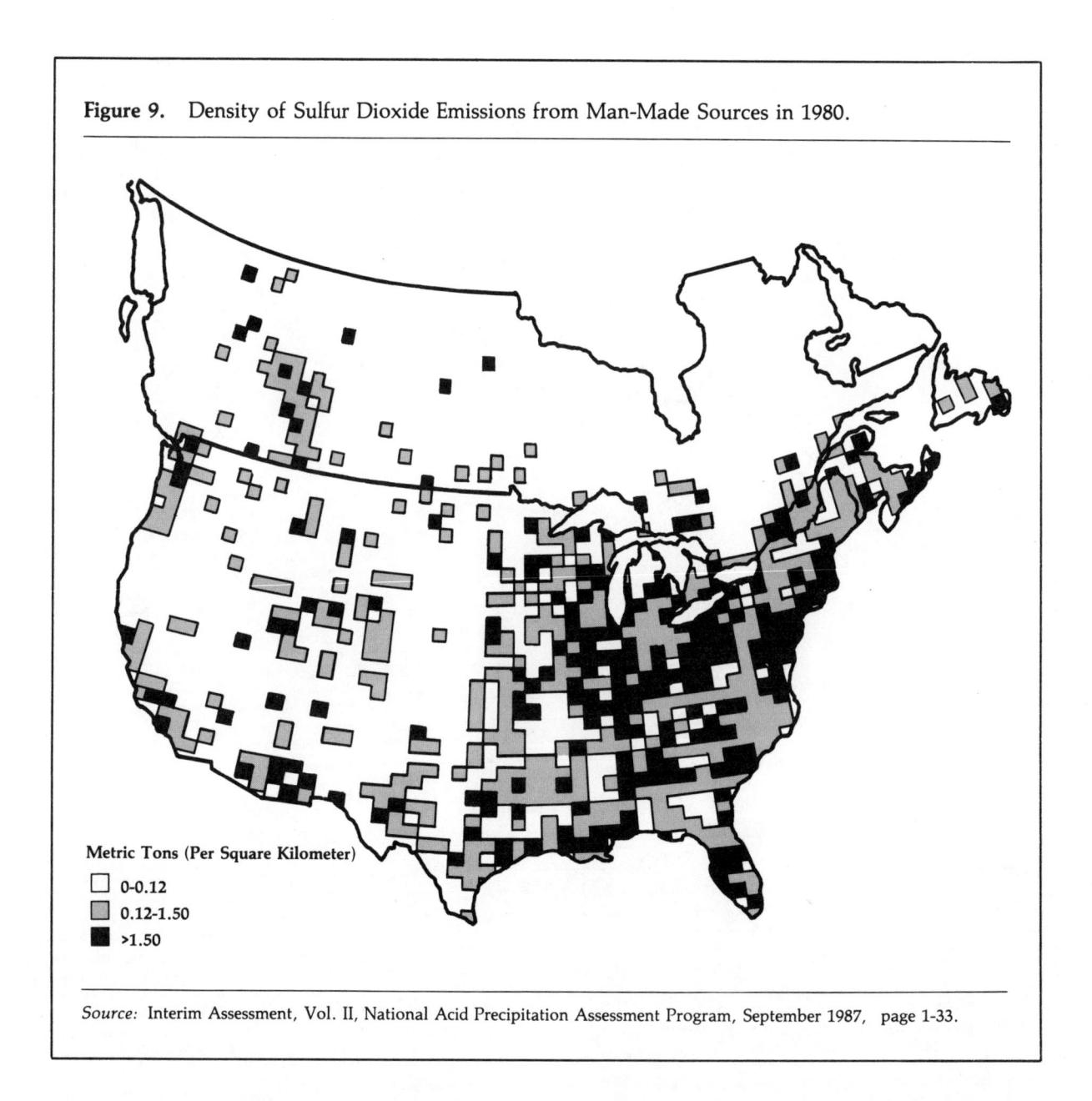

SO_2 is. Almost 70 percent of NO_x releases occur at elevations less than 120 feet.[185] For this reason, nitrogen oxides and associated ozone levels also tend to trouble cities more than sulfur oxides and their related conversion products do. Ultimately, though, nitrogen oxides are converted to nitric acid and, like sulfuric acid, can be deposited great distances from where they are emitted.

Table 6 lists the ten states with the highest NO_x emission levels, and Figure 11 displays regional emissions. High emission rates reflect high levels of fuel burning, especially in transportation and power generation. Nonetheless, compared to sulfur dioxide emissions, releases of nitrogen oxides are more dispersed. *(See Figure 9.)* Also of note: nitrogen oxide emissions tend to be constant year round since transpor-

Figure 10. Historical Trends in Nitrogen Oxide Emissions in the United States by Fuel Type.

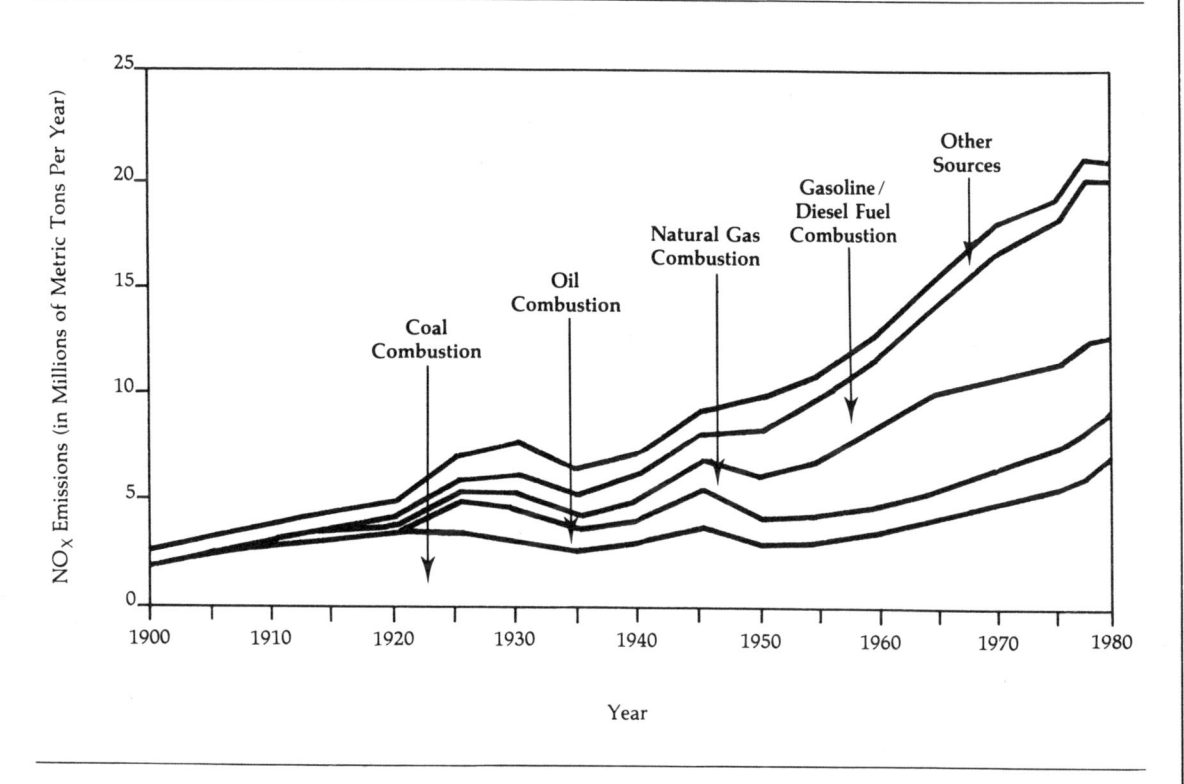

Source: Interim Assessment, Vol. II, National Acid Precipitation Assessment Program, September 1987, page 1-45.

tation and electric power-production activities vary little seasonally.

Volatile Organic Compounds. The term Volatile Organic Compounds (VOCs) covers a bewildering variety of chemicals that collectively play a critical role in ozone formation. Unlike oxides of sulfur and nitrogen, VOCs are emitted by all manner of activities. The largest single source of man-made VOCs is transportation, mostly from gasoline-powered vehicles, followed closely by industrial processes. *(See Table 7.)* Petroleum fuels and solvents are the two single largest sources of volatile hydrocarbons. Fuel combustion, primarily wood burning in homes, contributes about 12 percent to the total. Total anthropogenic VOC emissions rose between 1940 and 1970 and then gradually declined, thanks to automotive controls. Today's levels compare with those of 1950.

Natural sources in the United States, principally vegetation, also emit large amounts of organic compounds, perhaps more than man-made sources do. Approximately 90 percent of the natural VOCs are emitted from forests—60 percent from conifers and 30 percent from deciduous trees.[186] Summertime emissions from natural sources are greatest, estimated at 18 million tons. Total VOC emissions from natural

Table 6. Ten Leading Emitters Of NO_x in 1985 (Millions of Metric Tons of NO_2, and percent of total)

Texas	2.47	12.9%
California	1.11	5.8
Ohio	1.02	5.3
Pennsylvania	0.86	4.5
Illinois	0.81	4.2
Indiana	0.71	3.7
Florida	0.66	3.5
New York	0.65	3.4
Michigan	0.62	3.2
Louisiana	0.60	3.1

TOTAL: 9.51 Million Metric Tons (50% of total emissions)

Source: NAPAP Interim Assessment, Vol. II, pp. 1–52

Table 7. Man-Made Sources of VOCs in 1980) (Millions of Metric Tons, and percent of total)

Transportation	8.2	38.3%
Industrial processes	5.8	27.1
Fuel Combustion	2.5	11.7
Misc. Solvents	1.9	8.9
Petroleum storage & Marketing	1.4	6.5
Other	1.6	7.5
TOTAL	21.4	100%

Source: NAPAP Interim Assessment, Vol. II, pp. 1–58

sources in the United States could total 35 million tons annually, compared with 21 million tons from man-made sources.

As Figure 12 shows, man-made emissions are generally greatest in the eastern United States. The 11 states bordering Canada and the Great Lakes accounted for about 30 percent of total anthropogenic VOC emissions in 1980. Significantly, the south central states and the Southeast are major sources of both anthropogenic and natural VOCs. Of the eleven states that produce the most anthropogenic VOCs, nine also number among the largest emitters of nitrogen oxides. (*Compare Tables 8 and 6.*)

Formation of Secondary Pollutants

The pollutants of primary concern in forest decline and crop damage are formed in the atmosphere following the release of NO_x, VOCs, and SO_2. These secondary pollutants—nitric acid (HNO_3), ozone (O_3), and sulfuric acid (H_2SO_4)—are believed to be important contributors to crop and tree damages.

Ozone and Nitric Acid Formation. Ozone (O_3) occurs naturally in the atmosphere at relatively low levels and arises from several processes. An important source is the stratosphere, where oxygen atoms, formed when ultraviolet radia-

Table 8. Eleven Leading Emitters of Man-Made VOCs in 1980 (Millions of Metric Tons, and percent of total)

California	2.01	9.6%
Texas	1.97	9.4
New York	1.05	5.0
Pennsylvania	0.95	4.5
Ohio	0.91	4.3
Illinois	0.90	4.3
Michigan	0.83	4.0
Florida	0.75	3.6
New Jersey	0.63	3.0
Louisiana	0.57	2.7
Missouri	0.57	2.7

TOTAL: 11.14 Million Metric Tons (53% of total emissions)

Source: NAPAP Interim Assessment, Vol. II, pp. 1–62

Figure 11. Density of Nitrogen Oxide Emissions from Man-Made Sources in 1980.

Metric Tons (Per Square Kilometer)
- ☐ 0-0.12
- ▦ 0.12-1.50
- ■ >1.50

Source: Interim Assessment, Vol. II, National Acid Precipitation Assessment Program, September 1987, page 1-61.

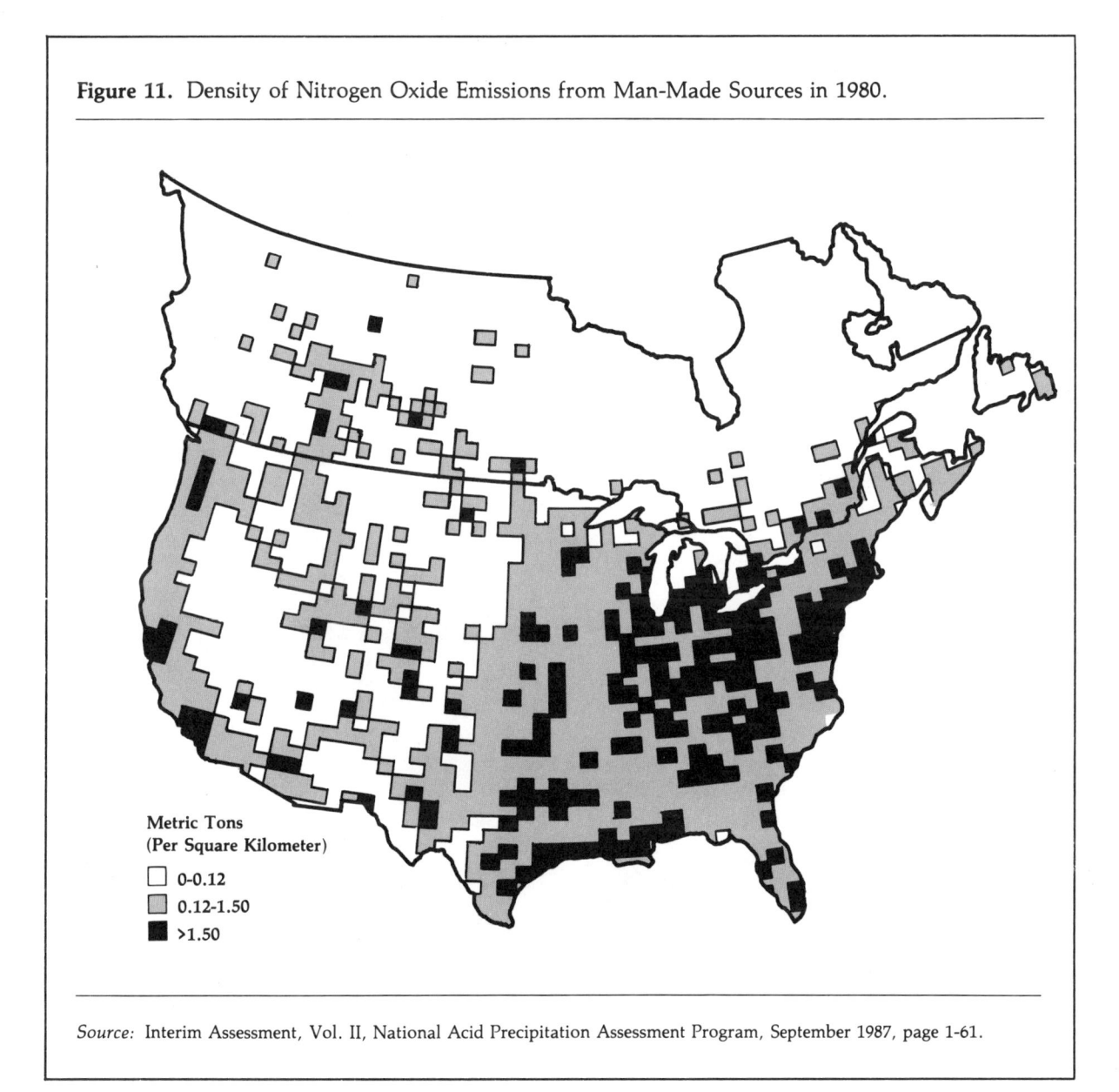

tion breaks up oxygen molecules, combine with molecular oxygen to form ozone. This stratospheric ozone is "good" ozone because it protects life on earth from the sun's intense ultraviolet radiation. But some finds its way to lower altitudes, where it contributes to background ozone levels of about 10–40 parts per billion.[187]

Ozone is also formed in the lower atmosphere through chemical reactions involving sunlight, oxygen, and hydrocarbons and NO_x from both natural and anthropogenic sources. This well-understood cycle begins when ultraviolet solar radiation breaks up nitrogen dioxide:

Figure 12. Density of Volatile Organic Compound Emissions from Man-Made Sources in 1980.

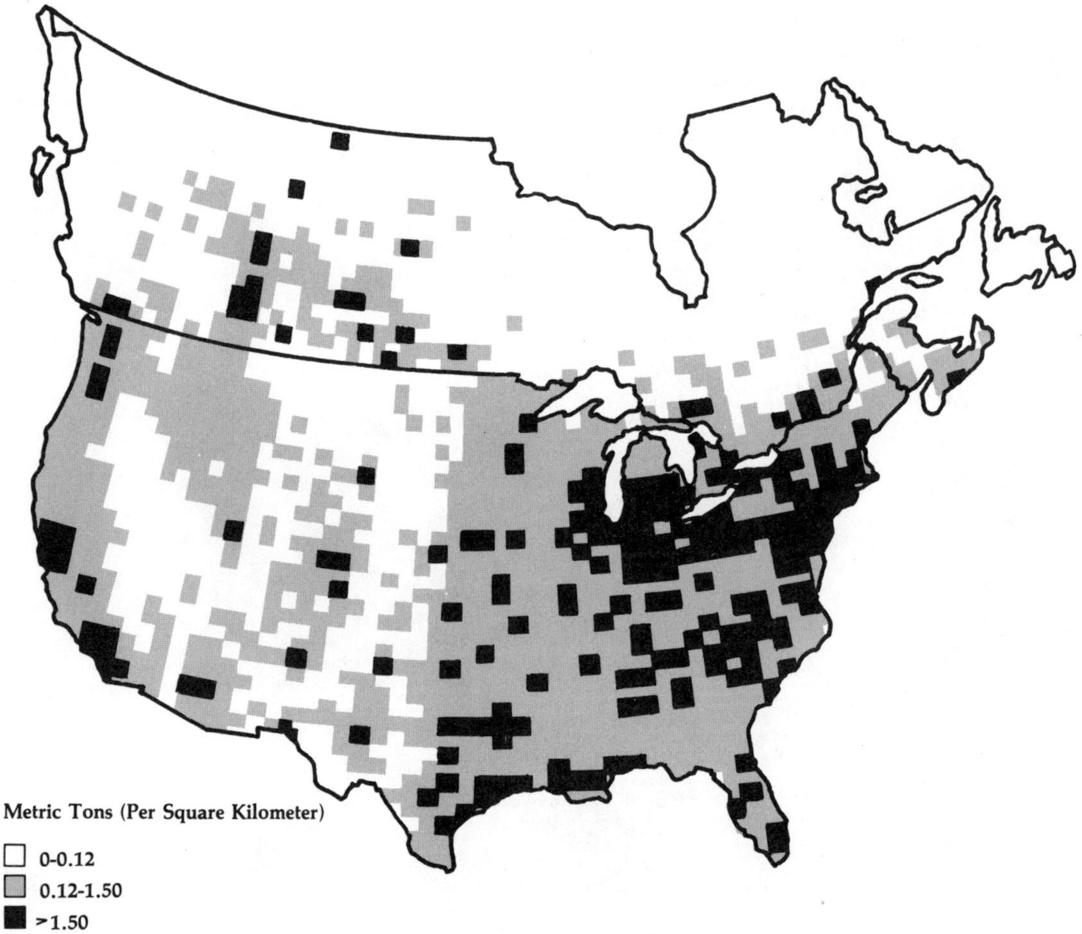

Metric Tons (Per Square Kilometer)
- ☐ 0–0.12
- ▨ 0.12–1.50
- ■ >1.50

Source: Interim Assessment, Vol. II, National Acid Precipitation Assessment Program, September 1987, page 1-61.

$$NO_2 + \text{solar radiation} \rightarrow NO + O$$

The single oxygen atom then reacts with an oxygen molecule to give ozone:

$$O_2 + O \rightarrow O_3$$

The cycle is completed when the ozone reacts with the NO, yielding nitrogen dioxide and oxygen.

$$NO + O_3 \rightarrow NO_2 + O_2$$

If there are constant concentrations of NO_2 and NO in the atmosphere, the ozone level is proportional to the *ratio* of NO_2 to NO. If this ratio is increased—through, say, the action of hydrocarbons—so is the level of ozone in the lower atmosphere ("bad" ozone because of its many ill effects on health and welfare).

Fuel burning results primarily in the formation of nitric oxide (NO) rather than nitrogen dioxide (NO_2). Unless some atmospheric NO is converted to NO_2 (increasing the ratio $[NO_2]/[NO]$), additional ozone will not form. It is here that non-methane volatile organic compounds (VOCs) and carbon monoxide (CO) play critical roles. Through complex chemical reactions, VOCs and CO help convert NO to NO_2, thereby increasing the ratio of NO_2 to NO and allowing ozone levels to increase. *(See Box.)*

The complex reactions creating ozone continue through the day, leading to higher levels of NO_2 (the reddish-brown gas commonly seen over urban areas) and ozone. Gradually, the NO_2 is removed from the atmosphere, primarily through the formation of nitric acid, an important element of acid deposition:

$$OH + NO_2 \rightarrow HNO_3 \text{ (nitric acid)}$$

The reactions involving NO_x occur rather quickly. In summer, NO_x has a lifetime of less than a day; in winter, about a week. Nitric acid is removed from the atmosphere through precipitation or surface deposition over one to ten days.[188] Ozone is removed from the lower atmosphere (the troposphere) as it is deposited on various surfaces and then broken up by sunlight, with the free oxygen atom joining with water to form hydroxyl radicals (OH).[189] During summer, ozone lasts about two days; in winter, about a month.

Sulfuric Acid Formation. Sulfuric acid (H_2SO_4) forms as sulfur dioxide (SO_2) oxidizes. This conversion can occur either in gas reactions or in liquids. The resulting sulfuric acid is removed either in precipitation or by dry deposition. An estimated 20 percent of the SO_2 emitted in the eastern United States and Canada is deposited in precipitation over this area; the rest is either deposited dry over land or else carried out over the Atlantic Ocean.[190] Since dry-deposited sulfur compounds eventually end up as sulfuric acid, estimating total acid deposition rates involves great uncertainties. Undoubtedly, estimates based soley on

---

One Way Ozone Forms

As an important example of ozone formation, consider the reactions involving carbon monoxide. Carbon monoxide first reacts with the hydroxyl radical (OH), which is usually present during the daytime, to form carbon dioxide and a hydrogen atom:

$$OH + CO \rightarrow CO_2 + H$$

The hydrogen then reacts with oxygen to form the hydroperoxyl radical, HO_2:

$$H + O_2 \rightarrow HO_2$$

The hydroperoxyl radical then converts the NO to NO_2, recycling the hydroxyl radical:

$$HO_2 + NO \rightarrow NO_2 + OH$$

The net result of these reactions is that carbon monoxide is converted to carbon dioxide and NO is converted to NO_2 with one oxygen molecule being consumed in the process.

---

wet acid deposition (rain and snow) significantly understate the total deposition rate.

The gas-phase conversion of SO_2 to sulfuric acid occurs by means of the hydroxyl radical (OH):

$$OH + SO_2 \rightarrow HSO_3$$
$$HSO_3 + O_2 \rightarrow HO_2 + SO_3$$
$$SO_3 + H_2O \rightarrow H_2SO_4$$

The hydroperoxyl radical (HO_2) is converted back to a hydroxyl radical (OH) through the conversion of NO to NO_2.[191] Thus, as long as nitric oxide (NO) is present in the atmosphere, sulfuric acid can be formed without consuming the hydroxyl radical, OH. This implies a linearity in gas-phase reactions between reductions in sulfur dioxide emissions and resulting reductions in sulfuric acid production: in short, reducing sulfur dioxide emissions by X percent will reduce sulfuric acid formation by the same percentage.

Reducing sulfur dioxide emissions by X percent will reduce sulfuric acid formation by the same percentage.

The conversion of SO_2 to sulfuric acid via gas-phase reactions is slow, typically consuming only 12 to 16 percent of the sulfur dioxide per day during summer and even less in winter when OH levels are lower. In the absence of precipitation, therefore, sulfuric acid can be formed from SO_2 long distances downwind from the SO_2 emission sources.

In clouds and rain, the conversion of SO_2 to sulfuric acid can be much quicker, up to 100 percent in an hour.[192] Two agents, ozone (O_3) and hydrogen peroxide (H_2O_2), are primarily responsible for the reactions, with highly water-soluble hydrogen peroxide considered the more important of the two. Hydrogen peroxide forms when two hydroperoxyl radicals (HO_2) react. Hydroperoxyl radicals form largely from reactions between carbon monoxide (CO) and the hydroxyl radical (OH). Once again, carbon monoxide's role in acid-deposition formation is important, though it defies intuition.

Seasonal Variations in Emissions and Wet Deposition. The seasonal emission and deposition patterns of the principal pollutants of concern here differ distinctly from one another. For example, emissions of sulfur dioxide (SO_2) and NO_x stay virtually constant throughout the year.[193] On the other hand, emissions of total VOCs (both natural and anthropogenic) are three times larger in summer than in winter. And, of course, more sunlight and warmer temperatures occur during the summer to drive photochemical, ozone-forming reactions. Not surprisingly, average summer concentrations are about twice those found in winter and account for the relatively higher *wet* deposition rates of sulfuric and nitric acid in summer compared to winter. Wet sulfate deposition in winter is often less than half and in some areas less than a fourth of corresponding summer values.[194] Nonetheless, total sulfur deposition during winter is not necessarily less. As indicated, what goes up must eventually come down, and the sulfur will come down either over land or the Atlantic Ocean.

Great uncertainties exist today in estimating the dry deposition rates of sulfur compounds. Some researchers believe that, on an annual average, wet and dry deposition rates are about equal.[195] If so, dry deposition would contribute significantly to the overall acid deposition problem.[196] Until scientists' understanding of dry deposition processes improves, determining the relative contributions of acidic dry and wet deposition will be difficult.

Regional Patterns of Pollutant Concentrations and Deposition

The emission patterns of the principal precursor air pollutants associated with ozone and acid deposition are shown in Figures 9, 11, and 12. But how do the patterns of pollutant concentrations and wet deposition rates at both low and high elevations correlate with observed forest and crop damages?

Acid Deposition. The acidity of precipitation at low elevations in the eastern United States is about ten times greater than would be expected for rain falling through unpolluted air. As Figure 13 shows, even at low elevations the lowest pH values (about pH 4.2) occur in the Northeast. At high elevations in the eastern half of the country, much of the moisture is deposited by clouds and the acidity is much greater. The average pH value for cloudwater at Mt. Mitchell in 1986 was only 3.3, more than ten times as acidic as the precipitation at low elevations. Some samples were as low as 2.3.[197] High acidity also characterizes the other eastern mountains where spruce and fir are declining. Clearly, in the high-altitude environments

Figure 13. Acidity (pH) of Wet Deposition for 1986.

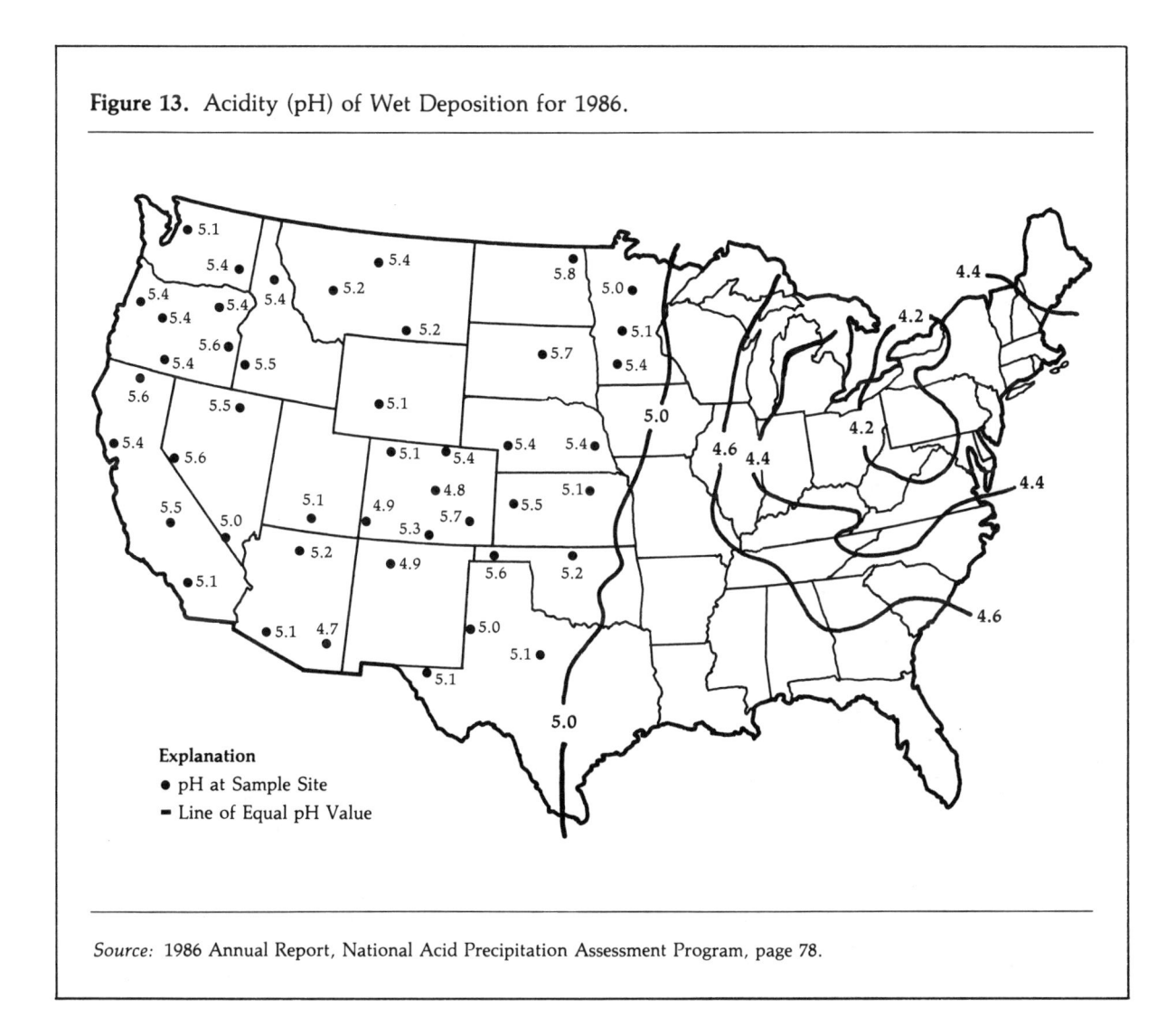

Source: 1986 Annual Report, National Acid Precipitation Assessment Program, page 78.

where trees are dying, moisture is much more acidic than at lower elevations.

Similar observations hold true for total wet acid deposition. As Figure 14 shows, annual wet acidic deposition values at low elevations in the eastern United States range from 0.3 to 0.6 kilograms of H^+ per hectare. The pattern resembles that for pH values. *(Compare Figures 13 and 14.)* At high elevations, where red spruce are dying, hydrogen deposition is much greater.

On Mt. Mitchell, annual H^+ deposition ranges between 2.0 and 4.6 kilograms per hectare, about ten times the deposition rate at low elevations.[198] By either measure—pH of moisture or total acid deposition—the mountains where forests are declining receive exceedingly high doses of acidic moisture.

Ozone Concentrations. Although ozone is not formed and transported like acid substances, the patterns of high exposures where trees are dying are similar. Urban areas have numerous

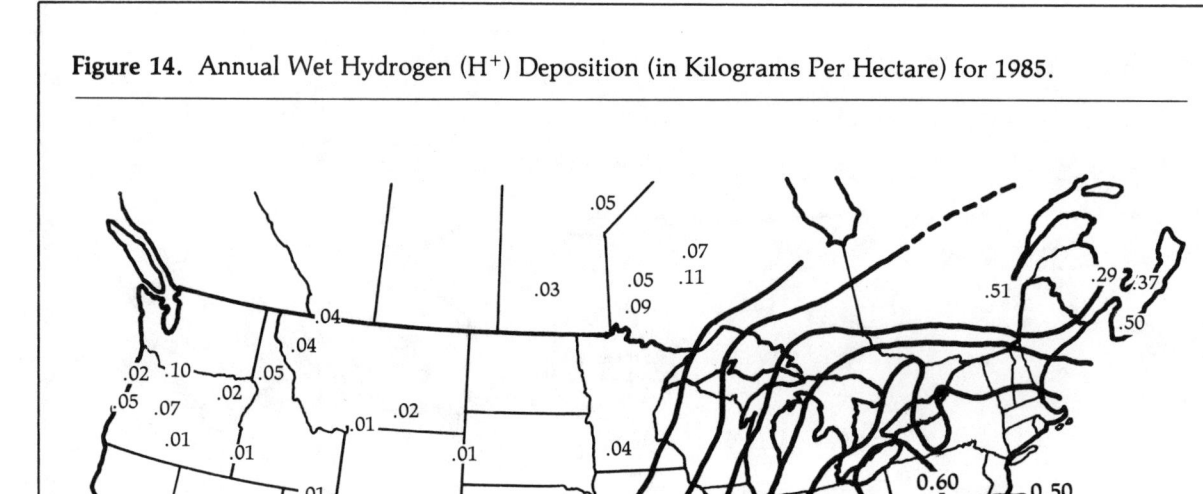

Figure 14. Annual Wet Hydrogen (H$^+$) Deposition (in Kilograms Per Hectare) for 1985.

Source: Interim Assessment, Vol. III, National Acid Precipitation Assessment Program, September 1987, page 5-42.

Acid "concentration" is measured by the pH of the moisture, a measure of the amount of hydrogen ion (H$^+$) in a liter of rainwater. In contrast, acid "deposition" takes into account the total amount of moisture actually falling on a region. Thus, deposition is measured as the total amount of hydrogen ions annually deposited over a given area.

sources of ozone precursors—NO$_x$, CO, and VOCs—so in the presence of sunlight, ozone forms readily. These precursors and the ozone they help create can travel great distances, threatening crops and materials hundreds of miles from the pollution source.[199] Pollution from the San Francisco Bay area, for example, has been traced 300 km (160 miles) to Yosemite National Park.[200] The Los Angeles plume has been followed for 350 km (190 miles) eastward across the Mohave Desert to the Colorado River.

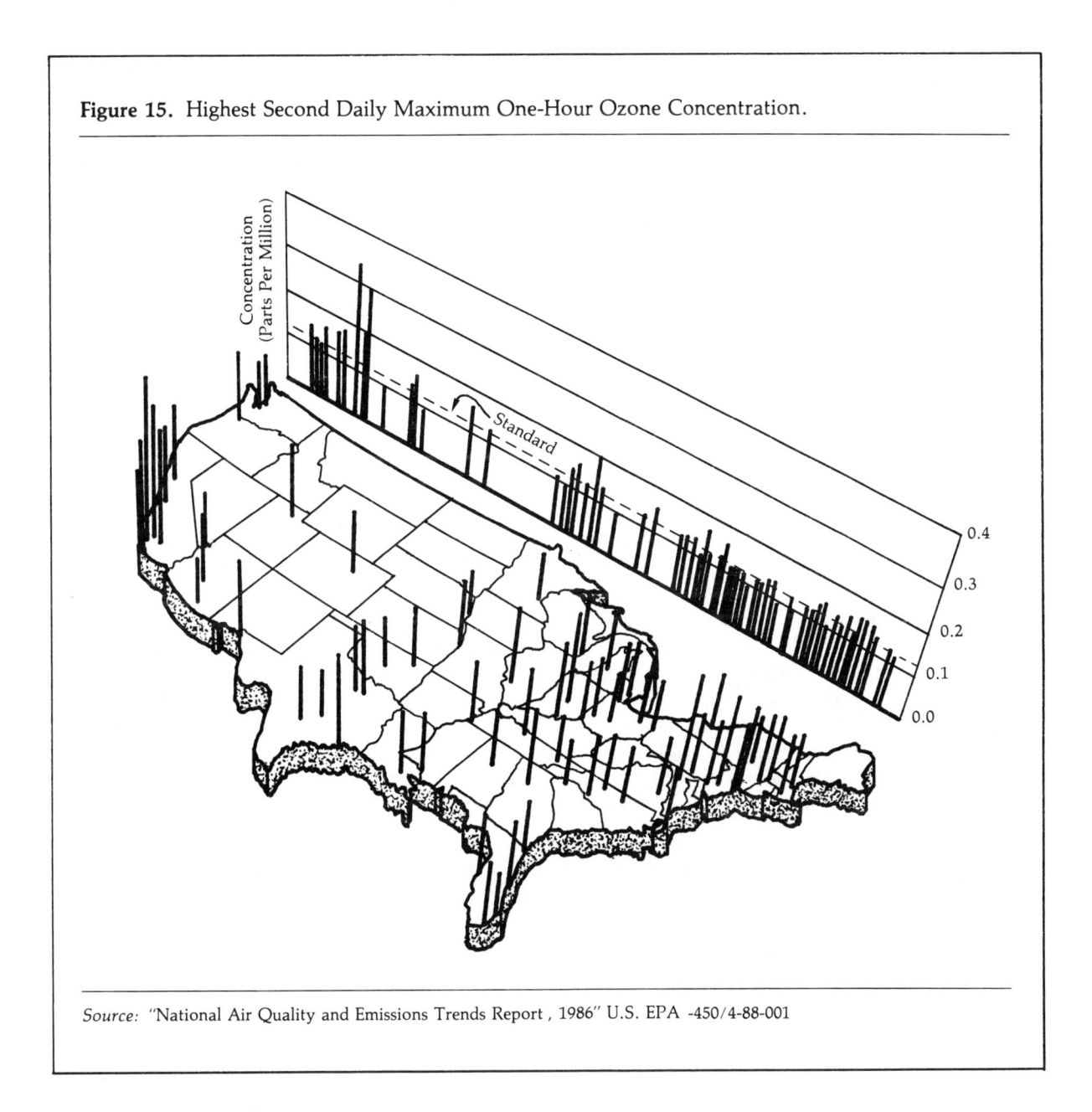

Figure 15. Highest Second Daily Maximum One-Hour Ozone Concentration.

Source: "National Air Quality and Emissions Trends Report, 1986" U.S. EPA -450/4-88-001

In rural areas that are not downwind from population centers, substantial ozone levels have also been measured. Recent research suggests that naturally occurring hydrocarbons, primarily from trees, in the presence of almost trace amounts of NO_x (about 1 ppb) can produce fairly large amounts of ozone.[201] As Figure 3 shows, average *rural* ozone levels in the United States are relatively uniform, which strongly suggests that unless NO_x emissions are significantly reduced, rural ozone and nitric acid deposition levels cannot be expected to decline.

Since short-term, high-dose ozone exposures cause more damage to crops and trees than low chronic levels do,[202] it is particularly useful to look at patterns of high exposures. As Figure 15 shows, in the eastern United States, peak ozone concentrations (measured mostly in urban areas) are comparable, in the range of 100 to 150 parts per billion (ppb). The average values for the second highest concentration vary from 130 ppb for low-population areas to 170 ppb for large metropolitan areas.[203] (The federal ozone standard is 120 ppb.) Ozone levels on remote eastern mountain tops are lower than these peak values, but nonetheless high. Ozone levels at Mt. Mitchell in North Carolina reached 112 ppb in 1986; at Whitetop Mountain in Virginia, they reached 120 ppb from October 1985 to September 1986.[204] Average ozone levels on Whiteface Mountain (New York) have been significantly higher than at other forested sites in Northern New England/ New York or New York/Pennsylvania/Maryland.[205] For example, between 1978 and 1983 the one-hour ozone concentration exceeded 70 ppb at Whiteface Mountain for 455 hours, while in Northern New England/NY and New York/Pennsylvania/Maryland this level was exceeded for only 103 and 223 hours, respectively. In the northeastern United States, only the New Jersey pinelands saw ozone levels exceeding 70 ppb more often than at Whiteface Mountain, 500 hours.

Patterns of Emissions and Patterns of Damages

Figure 16 combines the information on forest and crop damages from Section II *(Figure 1)* with the data on pollutant emissions from this section. Several patterns emerge. First, crop and forest damages are occurring throughout the Midwest and on both coasts. Second, significant losses in agricultural productivity are occurring in or near states that produce large amounts of ozone precursors (NO_x and VOCs). Third, states where forest declines or anomalous growth reductions in trees are significant lie downwind (generally east or northeast) of states with large emissions of sulfur dioxide and/or ozone precursors.

This geographical correlation does not, of course, establish cause and effect. However, combined with the extensive scientific evidence reviewed in Sections III and IV on damage mechanisms from air pollution, it further supports the contention that ozone and acid deposition are contributing to U.S. crop losses and forest declines.

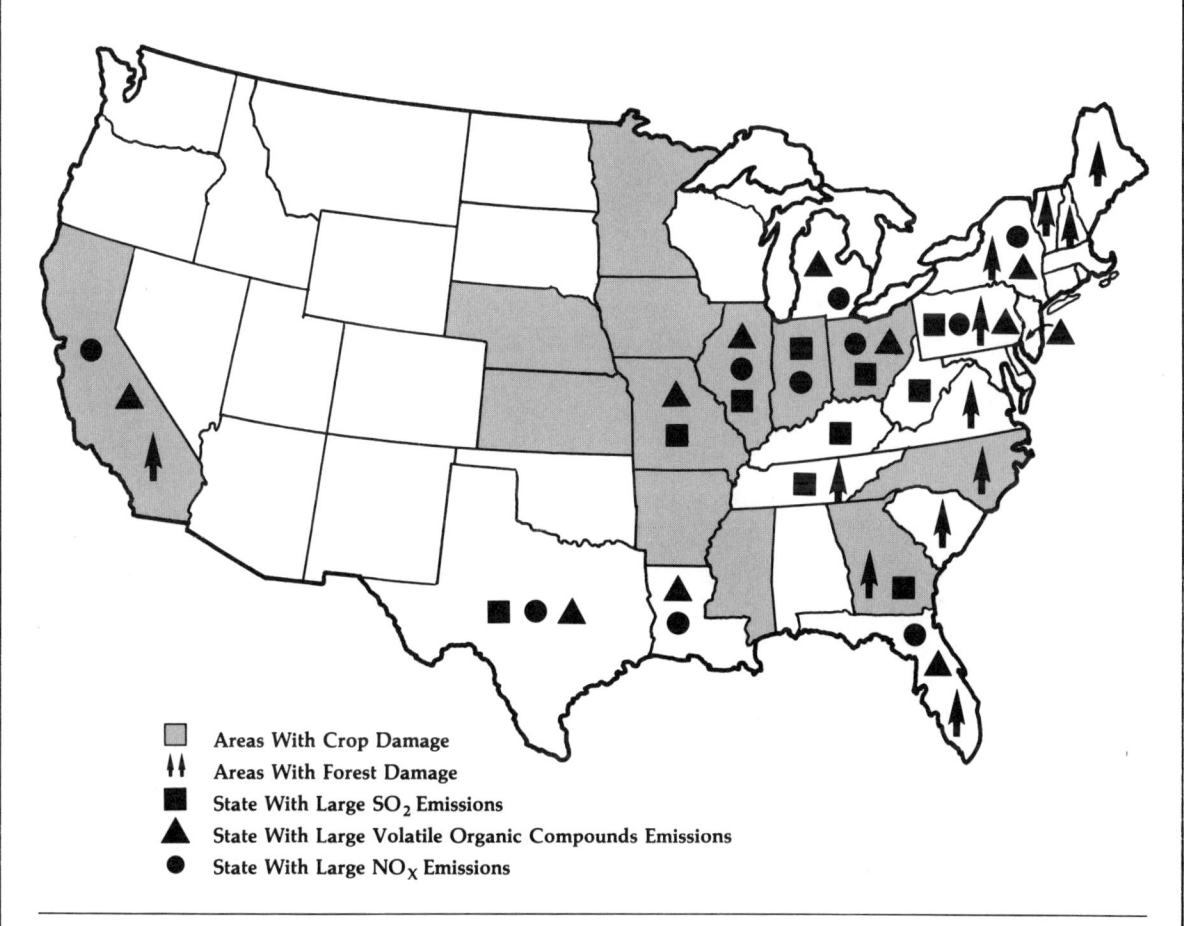

Figure 16. Comparison of States Where Forests and Crops are Being Damaged with Those Where Air Pollution Emissions are High.

- ▢ Areas With Crop Damage
- ↑↑ Areas With Forest Damage
- ■ State With Large SO_2 Emissions
- ▲ State With Large Volatile Organic Compounds Emissions
- ● State With Large NO_X Emissions

Source: The Authors

VI. Summary and Policy Recommendations

Summary of Pollution Damages to Forests and Crops

Across the United States, air pollution is contributing to the decline and ultimate death of forest trees and to widespread losses in crop yields. Extensive mortality caused by ozone air pollution is occurring among ponderosa and Jeffrey pines in southern California. Growth reduction and visible damage are taking a toll among eastern white pines—also the result of ozone pollution. And there is growing evidence that acid deposition and ozone are important contributors to the declines of several other stricken tree species in the East, including the high-elevation red spruce in the Appalachian Mountains from Vermont to North Carolina, and Fraser fir in the Southeast. Air pollution is also suspected as a factor in the decline of sugar maple and beech in Vermont's mountains and in the reduced growth (a common sign of ozone injury) of commercial yellow pines in much of the Southeast.

Trees are not the only plants being injured. For such sensitive crops as kidney beans, peanuts, soybeans, cotton, and winter wheat, current ozone losses vary from 5 to 20 percent. Total agricultural benefits from halving ozone levels have been estimated as high as $5 billion a year.

As documented in this report, the levels of air pollution are high where tree and crop injuries are occurring, and various direct and indirect damage mechanisms have been identified by which air pollution injures vegetation. On the mountains where the trees are dying, the acidity of the cloud moisture is about 10 times greater than that at low elevations and about 100 times greater than that of unpolluted precipitation. Similarly, total wet acidic deposition at high elevations is about ten times greater than at nearby lower elevations. Ozone levels are also high both on these mountains and in agricultural areas.

While natural stresses may deliver the final blow, air pollution readies the ground (or the tree itself) for forest decline.

Scientific evidence clearly indicates that acid deposition and ozone can injure foliage directly. In addition, acids can leach nutrients from both trees and soils. As the trees weaken, they succumb to such natural stresses as insects, disease, drought, or frost that they might otherwise have withstood. Thus, while natural stresses may deliver the final blow, air pollution readies the ground (or the tree itself) for forest decline.

Air pollution can travel tens and even hundreds of miles before it is removed from the atmosphere through wet or dry deposition. Consequently, acid deposition and ozone have become national and even international problems as trees and crops are threatened far from the sources of emissions. An examination of the pattern of injury shows that agricultural losses are occurring in or near states that are large sources of ozone precursors (NO_x and VOCs) and that states where forest declines or anomalous growth reductions are significant lie downwind (generally east or northeast) of states with large emissions of sulfur dioxide and/or ozone precursors. *(See Figure 16.)*

Given the inherent complexity of forests and the lack of long-term monitoring and study in the United States, it is understandable that some of the more recent forest declines are still not fully understood. Still, a consensus is emerging that current injuries arise from multiple stresses acting together, some of which are natural (e.g., weather extremes and insects) and some man-made (air pollution). As a consequence, the processes of decline vary by region with local climate conditions, rates of pollutant deposition, soil composition, and biological factors.

Unlike a healthy ecosystem, which can normally weather bouts of fire and insects, a system exposed to chronic air pollution gradually loses diversity and, hence, the capacity to cope with changing natural stresses.

The significance for these forest systems goes far beyond the immediate death of the damaged trees. As air pollution's effects on the ecosystem become progressively more severe, the whole ecosystem can be expected to deteriorate. Unlike a healthy ecosystem, which can normally weather bouts of fire and insects, a system exposed to chronic air pollution gradually loses diversity and, hence, the capacity to cope with changing natural stresses. Some scientists believe that ecosystems chronically exposed to air pollution can eventually collapse, as is happening on Camels Hump in Vermont and Mt. Mitchell, North Carolina.

Of great concern is the possibility that the forest damages now occurring primarily at high elevations could spread to lower sites and to other tree species. This pattern has already emerged in Europe and could occur in the United States if, for example, long-term acidification robs the soils at lower elevations of important nutrients. According to researchers at Oak Ridge National Laboratory, about 40 percent of the soils in the eastern United States are susceptible to substantial nutrient leaching.

The pollutants most implicated in forest damage are sulfuric acid, nitric acid, and ozone. To crops, ozone and other oxidants pose the most serious threats. The precursors of these pollutants—sulfur dioxide (SO_2), nitrogen oxides (NO_x), and volatile organic compounds (VOCs)—are produced when fossil fuels are burned in power plants, industrial boilers, and vehicles. These same pollutants also take a toll on human health, acidify lakes and streams, degrade visibility, and damage various materials.

Economic Impacts of Air Pollutants

While the economic damage to forests from air pollution cannot yet be quantified, financial losses from reduced crop yields have been estimated carefully. And other studies have been completed on the economic impacts of adverse health effects and decreased visibility.

The most recent and complete analysis, based on research of the National Crop Loss Assessment Network, shows that ozone is causing extensive agricultural losses and that reducing current ozone levels by 40 percent would lead to a $3.5 billion (1987 dollars) annual benefit in increased yields for eight crops.[206] If all affected crops were considered, the benefits would, of course, be much larger.

As for health benefits, deciding how much money pollution reduction is worth is a difficult task. Most researchers estimate how much reducing pollution by X amount would reduce various health risks to an individual. They then extrapolate the estimated dollar values of these risk reductions to the total affected population.[207] According to one estimate of ozone's effects, controls strong enough to simply meet the current primary ozone standard would yield benefits of $2.6 billion a year nationwide.[208]

Visibility degradation from air pollution also has high costs. EPA estimates the benefits of improved visibility at $700 million per year in the East from meeting current standards for SO_2,[209] while other researchers estimate benefits from meeting current particulate and sulfur dioxide standards at $500 million per year just for California's four major air basins.[210]

Future Pollution Trends

Those who oppose further emission controls argue that SO_2 releases in the United States have dropped by about 25 percent since the Clean Air Act was enacted and will continue to decline as new and cleaner power plants and vehicles replace older ones. They conclude that acid deposition and ozone problems will eventually disappear without further measures to reduce emissions. What do future trends for the implicated pollutants suggest?

Many factors will influence future sulfur dioxide emissions. These include future demand for electricity, the sulfur content of the fuels burned, the kinds of technologies that power plants use, and, especially, the retirement age of coal plants built before the New Source Performance Standard (NSPS) took hold. While the National Coal Association (NCA) estimates that sulfur emissions will continue to decline, at least through 1990,[211] the Electric Power Research Institute (EPRI) recently concluded that total sulfur emissions nationwide are likely to remain constant or increase slightly through about 2010.[212] A similar conclusion was reached by the recent Interim Assessment of the National Acid Precipitation Assessment Program (NAPAP).[213] One of the most important factors in forecasting future emissions is the assumption about the retirement age of older plants: since the Clean Air Act imposes stricter SO_2 emission standards on new sources than on older ones, refurbishing aging facilities to extend their lifetimes to 60 years (at costs of less than 50 percent of a new generating facility) keeps plants exempt from the stricter NSPS regulations. Retirement age, therefore, will be an important factor in assessing when total sulfur emissions will begin to decline.

Without further regulatory controls, NO_x pollution from power plants is projected to increase through 2030, roughly doubling emissions over present values.

The trend is up for NO_x emissions too. Without further regulatory controls, NO_x pollution from power plants is projected to increase through 2030, roughly doubling emissions over present values.[214] *(See Figure 17.)* Non-utility emissions—including those from transportation, industry, residential and commercial heating, and waste incineration—could increase by 30 percent between 1990 and 2030 unless technologies, regulations, or both are improved.[215]

Like NO_x emissions, volatile organic compounds (VOCs) released mainly from transportation, industrial processes, and other dispersed activities have been declining, primarily because vehicle emissions have been cut. However, by 1990 such emissions are expected to begin rising again until by 2030 they are 25 percent above current levels.[216] Since vehicle releases are strictly regulated, future VOC emissions will increasingly be from sources outside of transportation.

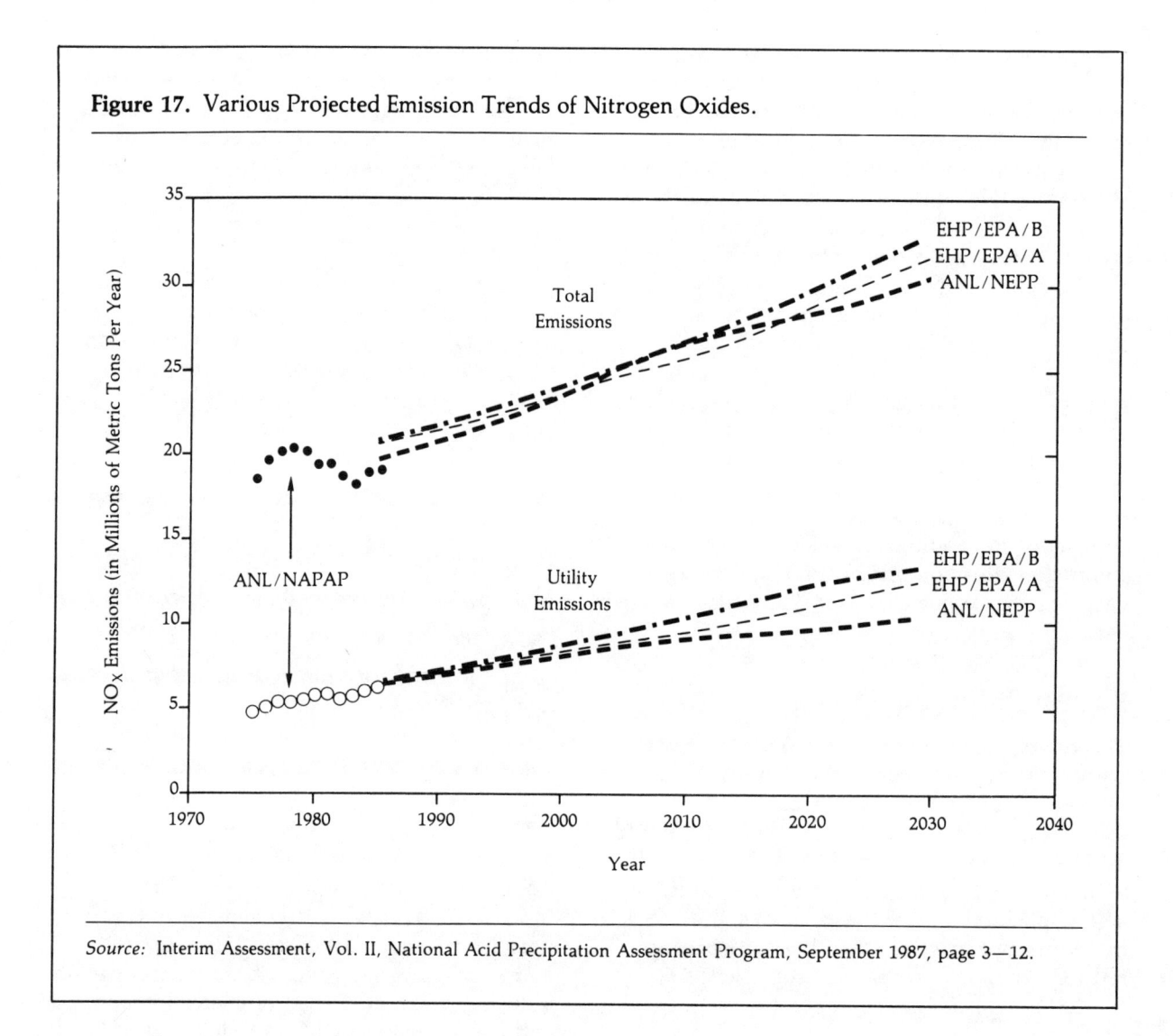

Figure 17. Various Projected Emission Trends of Nitrogen Oxides.

Source: Interim Assessment, Vol. II, National Acid Precipitation Assessment Program, September 1987, page 3—12.

These long-range projections, coupled with current evidence on the multiple adverse impacts of air pollution, make it clear that the nation's air pollution problems will worsen without new efforts to lower emissions. Unless pollutant levels are significantly reduced, we can only look forward to further forest damage, crop losses, ill health, water and soil acidification, damage to materials, and visibility losses.

Structural Shortcomings of the Clean Air Act

The Clean Air Act and its Amendments have failed to conquer air pollution partly because when the legislation was enacted in 1970 and 1977 understanding of how pollutants form, travel, and interact was limited. Ambient concentrations of such primary pollutants as sulfur dioxide and nitrogen oxides were then deemed the most relevant indices of the overall risks that these gases would pose. Too little consideration was given to the ultimate form of the compounds, air pollution's regional nature, the increased toxicity of the chemicals that primary pollutants eventually become, the long distances pollutants can travel before being deposited, or the cumulative impacts of acid deposition on soils and aquatic systems. The then-approved use of tall stacks to achieve local ambient air quality standards ultimately dispersed pollutants high into the atmosphere where they were transformed into acidic compounds and deposited many miles downwind.

*The Clean Air Act and its Amendments have failed to conquer air pollution partly because when the legislation was enacted in 1970 and 1977 understanding of how pollutants form, travel, and interact was limited. The Act also failed to anticipate how much the **number** of pollutant sources would grow.*

The Clean Air Act also failed to anticipate how much the *number* of pollutant sources would grow. The focus on sulfur fuel limits for individual boilers and emission limits for individual vehicles initially reduced pollution levels. But this approach proved limited as the number of cars, trucks, factories, and power plants increased with population and economic growth. Continuing this tack—focusing on technology-prescriptive regulatory limits on individual sources—will prove increasingly ineffective and costly.

Now is an opportune time to re-examine the basic approaches to reducing air pollution, bearing in mind several important facts. First, the costs of emission-control technologies in conventional vehicles and boilers can be expected to continue rising as ever more stringent limits are prescribed to meet ambient air quality standards. Second, long-term economic and population growth will increase the number of dispersed pollution sources—vehicles, small industrial facilities, homes, and commercial buildings—so the progress made by reducing emissions from individual sources will continue to erode. And, third, other pressing, energy-related national concerns discussed briefly below will have to be taken into account in developing pollution control strategies.

Air Pollution and Other National Issues

Air pollution problems (both the failure to meet urban air-quality goals and the long-range transport of airborne acids and ozone) are intimately linked to two other national issues, national oil security and greenhouse-induced climate change.[217] All three of these national concerns relate directly to patterns and rates of growth of fossil fuel combustion. As a result, long-term planning to control air pollution must take a broader perspective than simply reducing air pollution emissions. The connections among these issues are particularly significant in transportation (which depends almost totally on oil, ever more of which is imported) and in electric power generation (which is becoming more dependent on coal, a major source of air pollution and carbon dioxide).

A simple example makes these links clear. The automobile accounts for about 43 percent of total U.S. oil use. As such it contributes heavily to growing U.S. reliance on foreign oil that is imported increasingly from the Middle East. Cars are also a major source of carbon dioxide (CO_2) and ground-level ozone. Since these gases are both "greenhouse gases"—threatening long-term climate change—cars figure centrally in our climate problem. Finally, ground-level ozone in combination with other pollutants (including acid rain) poses risks to public health, crops, materials, and forests. Hence, cars also play an important part in health and environmental problems.

Clearly, public policies affecting air pollution must take these other considerations into account. If not, actions to mitigate one problem could easily neglect or even exacerbate others.

An Integrated Approach to Pollution Control

As with other environmental questions, the connections between air pollution, forest and crop damage, the erosion of health and materials, and the costs and benefits of corrective measures are not without uncertainty. But prudence and common sense dictate that, in a climate of scientific uncertainty, policies and regulatory actions should prevent harm while researchers strive for greater certainty. Doing nothing in the meantime would place sensitive ecosystems at the risk of irreversible damage.

Over the next ten to twenty years, air pollution control strategies should focus largely on reducing sulfur and nitrogen emissions from power plants, transportation systems, and other large sources of air pollutants. At the same time, the groundwork should be laid—through an ambitious research and development program—for the inevitable long-term transition to new energy sources that are inherently cleaner than the fossil fuels on which we now depend so heavily. A national strategy to reach these goals should have four major thrusts.

1. Controlling Stationary Source Emissions. Sulfur dioxide emissions must be cut substantially below present levels. A reasonable national goal for SO_2 emissions reduction is 10 million tons per year to be achieved in two phases, five million tons per year each, over a period of ten years. This cut represents, roughly, a 50-percent reduction from 1986 emissions—a goal consistent with the National Academy of Sciences' contention that a 50-percent reduction in acid (H^+) deposition would probably protect sensitive aquatic ecosystems.[218] Still, even a reduction of this magnitude may not protect all important resources. In 1984, the Office of Technology Assessment (OTA) concluded that wet sulfur deposition should be reduced by 50 to 80 percent in areas of high deposition and by 20 to 50 percent in other areas of eastern North America.[219]

Scientific uncertainty exists over the extent of NO_x reductions needed to protect health and other resources. NO_x emissions—three fourths of which derive from power plants and vehicles—lead to both ozone formation and to nitric acid deposition. Primarily as a result of fuel burning, global ozone levels in the lower atmosphere have approximately doubled over the past century. There can be little question that significant cuts in emissions of NO_x—the essential precursor of ozone—are needed. An initial reduction goal of five million tons per year to be achieved over the next ten years should be considered. This goal represents a 25-percent reduction from 1986 levels but a much larger reduction from projected growth in NO_x emissions.

The cost of modifying power plants to reduce emissions will be considerable—up to $7.3 billion a year, according to one estimate.[220] This estimate is based on the use of wet scrubbers, the most expensive, but proven means of SO_2 removal ($600 to $700 per ton of sulfur dioxide removed). The costs could be far lower if other technologies are used instead. For example, employing limestone injection multi-stage burners (LIMB) could bring the cost down to $400 to $500 per ton; with advanced slagging combustors, it would drop further to $50 to $100 per ton.[221]

To reduce compliance costs, a regulatory program must be flexible enough to permit creative combinations of measures to reach desired clean air goals. Overly prescriptive regulations, such as rules that in effect require wet scrubbers or that stipulate specific limits on the sulfur content of fuels, are unlikely to reduce emissions as efficiently or cost-effectively as possible.

One cost-effective strategy for reducing emissions from power plants—indeed, by far the cheapest means for reducing pollution—is to improve energy efficiency. In a study of the East Central Area Reliability (ECAR) power pool (including Ohio, Michigan, Indiana, Kentucky, West Virginia, and small sections of Maryland and Pennsylvania), the effects of various conservation options on emission levels from power plants were evaluated.[222] Options included more efficient residential appliances, electric motors, lighting, and measures to reduce heating and cooling needs in buildings. More efficient end-use technologies like these, the study found, could reduce electricity consumption in the ECAR region by 26 percent. Greater efficiency could, in turn, reduce SO_2 emissions by 7 to 11 percent during the 1990s. The study found that Midwestern consumers could realize a *net* savings of $4 to $8 billion if emission controls and conservation measures were both pursued, compared to a do-nothing scenario.

To encourage efficiency and other cost-effective options for reducing pollution, new regulations must allow full credit for the resulting pollutant decreases. One effective way is to prescribe pollutant caps for SO_2 and NO_x for each state and to allow the states (through their State Implementation Plans) flexibility in meeting the caps. States should be encouraged to consider various options to meet their goals: improved efficiency, fuel switching, retrofitting with clean coal technologies, transportation planning, and so forth. Establishing absolute caps on state-wide emissions would also force states to consider longer-term issues related to growth, land use, and local economics. Instead of relying solely on traditional rules and engineering fixes for pollution control, new regulations should include the use of incentives to reach the many dispersed sources of pollution and should allow emissions trading between sources within a state as well as between bordering states. New regulations could also rely on disincentives, such as a stiff tax on each ton of excess emissions beyond established caps. These taxes could be used to help defray compliance costs.

States should be allowed ten years to meet the two-phase emission reduction program. A decade provides a cushion against high initial costs and allows states time to develop innovative strategies to reach designated targets.

2. Reducing Vehicular Emissions. Transportation activities remain the single largest source of nitrogen oxides, volatile organic compounds, and carbon monoxide emissions. This is true even though auto-pollution controls have reduced emissions-per-vehicle-mile of nitrogen oxide, carbon monoxide, and hydrocarbon by 76, 96, and 96 percent, respectively, since 1968.[223] Unfortunately, increases in total vehicle miles traveled (VMT) have almost offset the reductions from individual autos. Partially as a result, over 80 million Americans live in areas where the ozone air standard is still exceeded, while 30 million live in areas where the carbon monoxide standard is routinely violated.

Over 80 million Americans live in areas where the ozone air standard is still exceeded, while 30 million live in areas where the carbon monoxide standard is routinely violated.

Losing the transportation battle for clean air is by no means inevitable. Transportation emissions can be cut significantly over the next 10 to 20 years through a combination of more stringent (but quite technologically feasible) emission limitations for cars, buses, and trucks; strengthened inspection and maintenance procedures; stronger measures to prevent tampering with pollution-control equipment; and the use of cleaner fuels (such as compressed natural gas) in commercial fleets and urban buses.

If the most stringent standards now proposed in Congress[224] were adopted along with better inspection and maintenance procedures, actual VOC emissions from mobile sources would drop by almost half by the year 2010 from 1985 values.[225] Without these measures,

VOC releases would increase slightly. With the proposed control effort, nitrogen oxide emissions would be cut by 30 percent by 2010 relative to 1985 values rather than increasing by about 15 percent, an overall reduction of 45 percent. The tighter emission limits (for VOC and NO$_x$) are already being met by about half of all new passenger cars and, if fully implemented, would cost roughly $50 to $125 per car.[226]

Programs that cut the total number of vehicle miles traveled and improve traffic flow would help reduce transportation emissions further. Promising measures include greater use of public transit; preferred parking spaces for car-and van-pools; the removal of subsidies for parking spaces except for car- and van-pools; provisions to encourage high-occupancy and non-motorized vehicles (bicycles) in all plans for new road construction; and the installation of dedicated lanes for high-occupancy vehicles on urban roadways.

Reducing the number of vehicle miles traveled would also slow the rate of climate change and enhance U.S. oil security. Today, transportation accounts for two thirds of domestic oil consumption. As a result of increasing oil demand (gasoline consumption grew 5 percent between 1985 and 1987), U.S. oil imports are also growing and now represent more than 40 percent of U.S. oil supply.

Improved vehicle fuel efficiency can be encouraged through various measures, including such regulatory requirements as the CAFE (Corporate Average Fuel Economy) standards, through financial tools (fuel taxes, graduated taxes on the purchase price of new vehicles according to efficiency, variable annual registration fees according to efficiency), and through federal and state government purchases of ultra-efficient vehicles.

3. Planning for the Long Term. Because air pollution is so intimately linked with energy use, long-term pollution-control planning must take into account two other national, energy-related problems: oil security and climate change. U.S. oil production in the lower-48 states peaked in 1970 and has been declining ever since. Prudhoe Bay production in Alaska is also nearing its peak. The depletion of U.S. oil resources and the increasing threat of climate change from the Greenhouse Effect are two long-term, and essentially irreversible, trends that will profoundly influence U.S. energy use and, hence, air pollution emissions.

Unless the United States is willing to import increasing amounts of oil from OPEC, it must sharply curb its petroleum consumption. Although the nation has huge coal resources and could make synthetic crude oil or methanol (to replace gasoline) from coal, such a policy would further increase carbon dioxide emissions[227] and accelerate climate change.

The policy implications for transportation planning are clear. Over the short term, national policies should emphasize more highly efficient conventionally powered vehicles with greatly reduced pollution emissions. At the same time, the development of electric and hydrogen-powered vehicles needs to be accelerated. Increased research and work on non-fossil fuel power sources for vehicles should be a top national priority. Introducing such inherently clean vehicles—with the hydrogen or electricity ultimately derived from non-fossil energy sources—would alleviate a whole range of national problems, including urban air pollution, acid deposition, climate change, and foreign oil dependency.

Electric power production is the other sector destined to see profound changes over the decades ahead. Driving these changes will be the need to reduce conventional pollutants and the risk of global climate change. To understand electricity's growing role in these issues, consider that in 1973, electricity generation accounted for 27 percent of U.S. primary fuel consumption and that, by 1986, U.S. power production accounted for 36 percent of primary energy consumption, about 70 percent of total SO$_2$ emissions, and a third of all U.S. carbon dioxide emissions. By 1986, some 70 percent of U.S. electricity was generated in fossil fuel plants, 55 percent in coal-fired plants alone. The Department of Energy expects a 50-percent increase in coal consumption between 1985 and the year 2000, with most of the coal being burned in electric power plants.[228]

Almost certainly, the new generation of clean coal technologies (integrated gasification combined cycle, fluidized bed combustion, in-duct sorbent injection, etc.) would allow power production with very few emissions of sulfur oxides, nitrogen oxides, or hydrocarbons. The use of these technologies would materially reduce the problem of acid deposition across the country. However, the problem of carbon dioxide emissions—the Greenhouse Effect— remains. Nothing now available or foreseeable can remove and dispose of the enormous quantities of carbon dioxide that coal-burning power plants would produce in the coming years. As a result, clean coal technologies can and should play an important, but ultimately transitional, role in our energy future.

Besides the evolution of clean new transportation technologies, the climate problem will also force the introduction of non-fossil fuel power generation. The renewable-energy technologies—solar cells, wind turbines, hydro power, geothermal energy—are strong candidates to assume the burden of future power production. The prospects of so-called Second Generation nuclear technologies (smaller, inherently safe, fuel-efficient, standardized fission reactors) offer a second, though less certain, option. In either case, air pollution as we know it today—high levels of ozone, acid deposition, carbon monoxide, and particulates—would be greatly reduced, along with the adverse effects now being visited on public health and welfare.

4. Improving Our Knowledge and Capabilities. The long-term ramifications of the forest declines occurring in the United States cannot be determined with great certainty until the mechanisms involved are better understood.

The federal government's principal investigation (the National Acid Precipitation Assessment Program) is scheduled to terminate in 1990. (The related federal program on crops, the National Crop Loss Assessment Network, ended in 1987.) It is highly unlikely that the unresolved scientific issues related to forest decline can be settled by 1990. Given the complexity and diversity of forest ecosystems, the host of possible contributing factors (pollution, weather extremes, biotic factors, and other anthropogenic causes), the importance of information for improving decision-making, and the prospect of long-term climate modification and elevated levels of ultraviolet radiation from stratospheric ozone depletion, a long-term research program must be maintained.

The National Acid Precipitation Assessment Program provides an umbrella mechanism for coordinating federal research in these areas. It should be reauthorized as the National Atmospheric Pollution Assessment Program and its scope expanded to support broad ecological research. Emphasis should be on gathering baseline data and understanding the ecological issues relevant to decision-makers.

As for air pollution's impacts on forest ecosystems, research should be expanded on both declining and apparently healthy forests to examine nutrient and pollutant flows, including those through soils. A number of representative forest ecosystems should be studied and monitored over time so that subtle changes can be detected early.

Long-term solutions to air pollution, acid precipitation, and greenhouse problems will ultimately require major innovations in energy technology. Since the need is pressing to move toward a more energy-efficient society powered primarily by non-fossil energy, research and development of energy technologies—particularly the renewable sources—should be given a very high priority. Accelerated development of transportation systems that do not depend on fossil fuels, such as electric or hydrogen-powered vehicles, should be supported. As these technologies become technically feasible, incentives for their introduction should be provided. One promising possibility is a tax on conventional transportation fuels; such a tax would reflect the enormous ecological, health, security, and climate risks not reflected in current fuel prices.

Research in energy storage also needs to be expanded. The absence of an inexpensive, light-weight, high energy-density battery is the principal obstacle to the widespread introduction of electric vehicles, a technology that could dramatically reduce air pollution emissions.

Improved methods for storing hydrogen should also be developed.

* * * * *

Finally, the need for top-level leadership to deal with the nation's interrelated air pollution, energy, and climate problems stands paramount. From the President to the Congress, our national leaders must recognize the immediacy of these interrelated problems and develop and implement cost-effective policies to protect valuable natural resources and sensitive ecosystems from permanent and irreversible damage.

James J. MacKenzie is a senior associate in World Resources Institute's Program in Climate, Energy, and Pollution. Formerly, Dr. MacKenzie was an energy policy analyst for the Union of Concerned Scientists. **Mohamed T. El-Ashry** is Vice President for Research and Policy Affairs at World Resources Institute. Previously, Dr. El-Ashry directed the environmental programs at the Tennessee Valley Authority.

Notes

1. W.H. Smith, *Air Pollution and Forests*, Springer-Verlag, New York, 1981; S.B. McLaughlin, "Effects of Air Pollution on Forests, A Critical Review," Journal of the Air Pollution Control Association, Vol. 35, No. 5, May 1985, pp. 512-534; F.H. Bormann, "Air Pollution and Forests: An Ecosystem Perspective," *BioScience*, Vol. 35, No. 7, July/August 1985, pp. 434-441; "Assessment of Crop Losses from Air Pollutants in the United States," Walter W. Heck, to be published in forthcoming WRI book on Multiple Air Pollutants.

2. "Multiple Pollutants and Forest Decline," Don Hinrichsen, *World Resources 1986*, published for the World Resources Institute by Basic Books; "Forest Damage and Air Pollution, Report of the 1986 Forest Damage Survey in Europe," International Cooperative Programme on Assessment and Monitoring of Air Pollution Effects on Forests, Global Environment Monitoring System, 1987; S. Nilsson and Peter Duinker, "The Extent of Forest Decline in Europe: A Synthesis of Survey Results," *Environment*, Vol. 29, No. 9, November 1987, pp. 4-31; R.F. Huettl, " 'New Type' of Forest Damages in Central Europe," in forthcoming WRI book on Multiple Air Pollutants, 1989.

3. Peter Schuett and Ellis B. Cowling, "Waldsterben, a General Decline of Forests in Central Europe: Symptoms, Development, and Possible Causes," *Plant Disease*, Vol. 69, No. 7, July 1985, pp. 548-558; R.F. Huettl, " 'New Type' of Forest Damages in Central Europe," in forthcoming WRI book on Multiple Air Pollutants, 1989; Bernhard Prinz "Causes of Forest Damage in Europe: Major Hypotheses and Factors," *Environment*, Vol. 29, No. 9, November 1987, pp. 10-37.

4. William H. Smith, "Forest Quality and Air Quality," *Journal of Forestry*, 83 No. 2 (February, 1985): 82-92.

5. Paul D. Manion, "Factors Contributing to the Decline of Forests, A Conceptual Overview," presented to the symposium on Air Pollutant Effects on Forest Ecosystems, St. Paul, Minnesota, May 8-9, 1985, 63-73.

6. Paul D. Manion, *Tree Disease Concepts* (Ingleside Cliffs, NJ: Prentice Hall Inc., 1981), 399.

7. Manion, "Factors Contributing to the Decline of Forests."

8. Philip M. Wargo, "Interaction of Stress and Secondary Organisms in Decline of Forest Trees," presented to the symposium on Air Pollutant Effects on Forest Ecosystems, St. Paul, Minnesota, May 8-9, 1985, 75-86; Paul R. Miller, "Concept of Forest Decline in Relation to Western Forests," in forthcoming WRI book on Multiple Air Pollutants, 1989.

9. Wargo, "Interaction of Stress and Secondary Organisms in Decline of Forest Trees."

10. U.S. Department of Agriculture, Forest Service, and Pennsylvania State University, Department of Plant Pathology, "Diagnosing Injury to Eastern Forest Trees," 1987.

11. Paul R. Miller, "Concept of Forest Decline in Relation to Western Forests," in forthcoming WRI book on Multiple Air Pollutants, 1989.

12. Gerard D. Hertel, U.S. Forest Service, "Spruce-Fir Research Cooperative Technical Report" March 15, 1988.

13. Ellis B. Cowling, "Comparison of Regional Declines of Forests in Europe and North America: The Possible Role of Airborne Chemicals," presented to the symposium on Air Pollutant Effects on Forest Ecosystems, St. Paul, Minnesota, May 8–9, 1985, 217–234.

14. Cowling, "Comparison of Regional Declines of Forests in Europe and North America."

15. Cowling, "Comparison of Regional Declines of Forests in Europe and North America," p. 222; J.N. Woodman, Ellis B. Cowling, "Airborne Chemicals and Forest Health," *Environ. Sci. Technol.*, Vol. 21, No. 2, 1987, pp. 122 ff.

16. National Acid Precipitation Assessment Program, *Interim Assessment, The Causes and Effects of Acid Deposition*, (Washington: Volume IV, Effects of Acidic Deposition, September, 1987), pp. 7–1 ff.

17. Miller, "Concept of Forest Decline in Relation to Western Forests."

18. Miller, "Concept of Forest Decline in Relation to Western Forests."

19. Symposium on Air Pollutants Effects on Forest Ecosystems, St. Paul, MN, May 8–9, 1985; A.H. Johnson and S.B. McLaughlin, "The Nature and Timing of the Deterioration of Red Spruce in the Northern Appalachian Mountains," *Acid Deposition Long-Term Trends*, (Washington: National Academy Press, 1986); National Acid Precipitation Assessment Program, *Interim Assessment, The Causes and Effects of Acid Deposition*, Volume IV.

20. National Acid Precipitation Assessment Program, *Interim Assessment, The Causes and Effects of Acid Deposition*, Volume IV, 7–12 ff; Miller, "Concept of Forest Decline in Relation to Western Forests."

21. U.S. Environmental Protection Agency, *Air Quality Criteria for Ozone and Other Photochemical Oxidants, Volume III of V* EPA/600/8–84/020cF (Washington D.C.: U.S. Government Printing Office, August 1986) 1986–646–116/40656, page 7–33.

22. Miller, "Concept of Forest Decline in Relation to Western Forests."

23. Miller, "Concept of Forest Decline in Relation to Western Forests."

24. J.N. Woodman, Ellis B. Cowling, "Airborne Chemicals and Forest Health," *Environ. Sci. Technol.*, Vol. 21, No. 2, 1987, p. 122.

25. National Acid Precipitation Assessment Program, *Interim Assessment, The Causes and Effects of Acid Deposition*, Volume IV, pp. 7–13.

26. National Acid Precipitation Assessment Program, *Interim Assessment, The Causes and Effects of Acid Deposition*, Volume IV, pp. 7–14.

27. National Acid Precipitation Assessment Program, *Interim Assessment, The Causes and Effects of Acid Deposition*, Volume IV, pp. 7–14.

28. National Acid Precipitation Assessment Program, *Interim Assessment, The Causes and Effects of Acid Deposition*, Volume IV, pp. 7–14.

29. For a detailed discussion on forest decline in Germany see *World Resources 1986*.

30. "Statement of Baron Franz Riederer Von Paar," in U.S. House Subcommittee on Mining, Forest Management, and Bonneville Power Administration of the Committee on Interior and Insular Affairs, *Effects of Air Pollution and Acid Rain on Forest Decline*, Hearings, June 7, 1984.

31. B. Prinz "Causes of Forest Damage in Europe," *Environment* 29 (November 1987): 11.

32. Ebasco Services Incorporated, "Acid Deposition Studies," prepared for the Business Roundtable Environment Task Force, (New York: Ebasco, 1986) I–11.

33. Huettl, "'New Type' of Forest Damages."

34. Huettl, "'New Type' of Forest Damages."

35. S. Nilsson and P. Duinker "The Extent of Forest Decline in Europe," *Environment* 29 (November 1987): 4.

36. Huettl, "'New Type' of Forest Damages."

37. Huettl, "'New Type' of Forest Damages."

38. Huettl, "'New Type' of Forest Damages."

39. Huettl, "'New Type' of Forest Damages."

40. Huettl, "'New Type' of Forest Damages."

41. Huettl, "'New Type' of Forest Damages."

42. Peter Schuett and Ellis B. Cowling, "Waldsterben, a General Decline of Forests in Central Europe: Symptoms, Development, and Possible Causes," *Plant Disease*, Vol. 69, No. 7, July 1985, p. 548 ff.

43. Huettl, "'New Type' of Forest Damages."

44. National Acid Precipitation Assessment Program, *Interim Assessment, The Causes and Effects of Acid Deposition*, Volume IV, pp. 7–14 ff.

45. Gerard D. Hertel, U.S. Forest Service, "Spruce-Fir Research Cooperative Technical Report," March 15, 1988.

46. Arthur H. Johnson and Thomas G. Siccama, "Decline of Red Spruce in the High-Elevation Forests of the Northeastern U.S.," in forthcoming WRI book on Multiple Air Pollutants, 1989.

47. Hertel, U.S. Forest Service, "Spruce-Fir Research Cooperative Technical Report," Arthur H. Johnson and Thomas G. Siccama, "Decline of Red Spruce," A.J. Friedland and J.J. Battles, "Red Spruce (*Picea Rubens* Sarg.) Decline in the Northwestern United States: Review and Recent Data From Whiteface Mountain" *Proceedings of the Workshop on Forest Decline and Reproduction: Regional and Global Consequences*, Krakow, Poland, March 23–27, 1987, L. Kairiukstis, S. Nilsson and A. Straszak (Eds.), 1987 IIASA, A-2361, Laxenburg, Austria.

48. Johnson and Siccama, "Decline of Red Spruce."

49. Hertel, U.S. Forest Service, "Spruce-Fir Research Cooperative Technical Report."

50. G.D. Hertel *et al.*, "Status of the Spruce-Fir Cooperative Research Program," presented at the 80th Annual Meeting of the Air Pollution Control Association, New York, June 21–26, 1987; National Acid Precipitation Assessment Program, *Interim Assessment, The Causes and Effects of Acid Deposition*, Volume IV, pp. 7–17.

51. Johnson and Siccama, "Decline of Red Spruce."

52. Johnson and Siccama, "Decline of Red Spruce."

53. Johnson and Siccama, "Decline of Red Spruce."

54. National Acid Precipitation Assessment Program, *Interim Assessment, The Causes and Effects of Acid Deposition*, Volume IV, pp. 7–15.

55. Johnson and Siccama, "Decline of Red Spruce."

56. Robert I. Bruck "Forest Decline in the Southeastern United States," in forthcoming WRI book on Multiple Air Pollutants, 1989.

57. Bruck, "Forest Decline in the Southeastern United States."

58. Bruck, "Forest Decline in the Southeastern United States."

59. Hertel, U.S. Forest Service, "Spruce-Fir Cooperative Technical Report."

60. Hertel, U.S. Forest Service, "Spruce-Fir Cooperative Technical Report."

61. National Acid Precipitation Assessment Program, *Interim Assessment, The Causes and Effects of Acid Deposition*, Volume IV, pp. 7–17.

62. R.M. Sheffield *et al.*, *Pine Growth Reductions in the Southeast*, Resource Bulletin SE-83 (Asheville, NC: Southeastern Forest Experiment Station, USDA, Forest Service, November 1985) pp. 1–2.

63. Sheffield *et al.*, *Pine Growth Reductions in the Southeast*, p. 6.

64. R.M. Sheffield and N.D. Cost, "Behind the Decline," *Journal of Forestry*, 85 No. 1 (January, 1987): 29.

65. Sheffield and Cost, "Behind the Decline."

66. Sheffield *et al.*, *Pine Growth Reductions in the Southeast*. p. 2.

67. National Acid Precipitation Assessment Program, *Interim Assessment, The Causes and Effects of Acid Deposition*, Volume IV, pp. 7–20.

68. Brent Mitchell "Air Pollution and Maple Decline," *Nexus* Vol. 9, No. 3 published by the Atlantic Center for the Environment, a division of the Quebec-Labrador Foundation, (Summer, 1987) p. 1.

69. National Acid Precipitation Assessment Program, *Interim Assessment, The Causes and Effects of Acid Deposition*, Volume IV, pp. 7–21.

70. Brent Mitchell "Air Pollution and Maple Decline," p. 2.

71. Brent Mitchell "Air Pollution and Maple Decline," p. 4.

72. Brent Mitchell "Air Pollution and Maple Decline," p. 4.

73. U.S Department of Commerce, *Statistical Abstract of the United States, 1987*, (Washington D.C.: U.S. Government Printing Office, 1986) p. 642.

74. Interagency Task Force on Acid Precipitation, *Report on the Crop Response Workshop*, held in Chicago, Illinois, April 17–18, 1986. p. 1.

75. National Acid Precipitation Assessment Program, *Interim Assessment, The Causes and Effects of Acid Deposition*, Volume IV, pp. 6–1.

76. National Acid Precipitation Assessment Program, *Interim Assessment, The Causes and Effects of Acid Deposition*, Volume IV, pp. 6–1.

77. Heck "Assessment of Crop Losses."

78. Heck "Assessment of Crop Losses."

79. W.W. Heck, W.W. Cure, D.S. Shriner, R.J. Olson, and A.S. Heagle, "Ozone Impacts on the Productivity of Selected Crops," in the Proceedings of the Symposium on the *Effects of Air Pollution on Farm Commodities*, February 18, 1982, Washington, D.C., Izaak Walton League of America, pp. 147–176.

80. Schuett and Cowling, "Waldsterben, A General Decline of Forests in Central Europe."

81. Johnson and Siccama, "Decline of Red Spruce."

82. National Acid Precipitation Assessment Program, *Interim Assessment, The Causes and Effects of Acid Deposition*, (Washington: Volume II, Emissions and Control, September, 1987), p. 1-25 ff.

83. Miller, "Concept of Forest Decline in Relation to Western Forests."

84. Miller, "Concept of Forest Decline in Relation to Western Forests."

85. U.S. Environmental Protection Agency, *Air Quality Criteria for Ozone and Other Photochemical Oxidants, Volume III of V* EPA/600/8-84/020cF (Washington D.C.: U.S. Government Printing Office, August 1986) 1986-646-116/40656, page 7-37.

86. Miller, "Concept of Forest Decline in Relation to Western Forests."

87. Miller, "Concept of Forest Decline in Relation to Western Forests."

88. Miller, "Concept of Forest Decline in Relation to Western Forests."

89. Miller, "Concept of Forest Decline in Relation to Western Forests."

90. Huettl, " 'New Type' of Forest Damages."

91. Schuett and Cowling, "Waldsterben, a General Decline of Forests in Central Europe,"; Bernhard Prinz, "Major Hypotheses and Factors, Causes of Forest Damage in Europe," Environment, Vol. 29, No. 9, November, 1987.

92. Schuett and Cowling, "Waldsterben, a General Decline of Forests in Central Europe."

93. Volker A. Mohnen, "Exposure of Forests to Gaseous Air Pollutants and Clouds," Mountain Cloud Chemistry Program, Report submitted to EPA, June 1987, Section 6.

94. Ellis B. Cowling, "Comparison of Regional Declines of Forests in Europe and North America: The Possible Role of Airborne Chemicals," presented to the symposium on Air Pollutant Effects on Forest Ecosystems, St. Paul, Minnesota, May 8-9, 1985, p. 221.

95. National Acid Precipitation Assessment Program, *Interim Assessment, The Causes and Effects of Acid Deposition*, Volume IV, pp. 7-15.

96. Bruck, "Forest Decline in the Southeastern United States."

97. National Acid Precipitation Assessment Program, *Interim Assessment, The Causes and Effects of Acid Deposition*, Volume IV, pp. 7-15.

98. National Acid Precipitation Assessment Program, *Interim Assessment, The Causes and Effects of Acid Deposition*, (Washington: Volume III, Atmospheric Processes, September, 1987), pp. 5-49.

99. Hertel, U.S. Forest Service, "Spruce-Fir Research Cooperative Technical Report."

100. National Acid Precipitation Assessment Program, *Interim Assessment, The Causes and Effects of Acid Deposition*, Volume III, pp. 5-42.

101. V.K. Saxena *et al.*, "Monitoring the Chemical Climate of the Mt. Mitchell State Park for Evaluating Its Impact on Forest Decline," 1988, Tellus, in press.

102. Bruck, "Forest Decline in the Southeastern United States."

103. Volker A. Mohnen, "Exposure of Forests to Gaseous Air Pollutants and Clouds," Chapter 6, Mountain Cloud Chemistry Program, June 1987.

104. Volker A. Mohnen, "Exposure of Forests to Gaseous Air Pollutants and Clouds."

105. National Acid Precipitation Assessment Program, *Interim Assessment, The Causes and Effects of Acid Deposition*, Volume IV, pp. 7-23.

106. Huettl, "'New Type' of Forest Damages."

107. National Acid Precipitation Assessment Program, *Interim Assessment, The Causes and Effects of Acid Deposition*, Volume IV, pp. 7-23.

108. U.S. Environmental Protection Agency, *Air Quality Criteria for Ozone and Other Photochemical Oxidants, Volume III of V* EPA/600/8-84/020cF (Washington D.C.: U.S. Government Printing Office, August 1986) 1986-646-116/40656, page 7-12.

109. U.S. Environmental Protection Agency, *Air Quality Criteria for Ozone and Other Photochemical Oxidants, Volume III of V*, p. 7-7.

110. U.S. Environmental Protection Agency, *Air Quality Criteria for Ozone and Other Photochemical Oxidants, Volume III of V* pp. 7-8.

111. A.H. Johnson and T.G. Siccama, "Acid Deposition and Forest Decline," *Environmental Science and Technology* 17, No. 7 (July 1983): 294A.

112. P.B. Reich and R.G. Amundson, "Ambient Levels of Ozone Reduce Net Photosynthesis in Tree and Crop Species, Science, 230: 566-570, 1985.

113. Reich and Amundson, "Ambient Levels of Ozone Reduce Net Photosynthesis in Tree and Crop Species."

114. National Acid Precipitation Assessment Program, *Interim Assessment, The Causes and Effects of Acid Deposition*, Volume IV, pp. 7-35.

115. Leonard H. Weinstein, personal communication, Feb. 1988.

116. U.S. Environmental Protection Agency, *Air Quality Criteria for Ozone and Other Photochemical Oxidants, Volume III of V*, pp. 7-15.

117. Huettl, "'New Type' of Forest Damages."

118. National Acid Precipitation Assessment Program, *Interim Assessment, The Causes and Effects of Acid Deposition*, Volume IV, pp. 7-28.

119. J.S. Jacobson and J.P. Lassoie, "Response of Red Spruce to Sulfur- and Nitrogen-Containing Contaminants in Simulated Acidic Mist," to be published in the Proceedings of the Symposium on the Effects of Atmospheric Pollution on Spruce and Fir Forests in the Eastern United States and the Federal Republic of Germany, October 18-23, 1987, Burlington, Vermont.

120. National Acid Precipitation Assessment Program, *Interim Assessment, The Causes and Effects of Acid Deposition*, Volume IV, pp. 7-28.

121. National Acid Precipitation Assessment Program, *Interim Assessment, The Causes and Effects of Acid Deposition*, Volume IV, pp. 7-29.

122. National Acid Precipitation Assessment Program, *Interim Assessment, The Causes and Effects of Acid Deposition*, Volume IV, pp. 7-29.

123. National Acid Precipitation Assessment Program, *Interim Assessment, The Causes and Effects of Acid Deposition*, Volume IV, pp. 7-29.

124. Bruck, "Forest Decline in the Southeastern United States."

125. C. Rinalla *et al.* "Effects of Simulated Acid Deposition on the Surface Structure of Norway Spruce and Silver Fir Needles,"

Eur. J. For. Path. 16: 440–446, 1986, (as cited in Bruck, 1989).

126. Bruck, "Forest Decline in the Southeastern United States."

127. V.K. Saxena et al. "Monitoring the Chemical Climate of the Mt. Mitchell State Park."

128. A.J. Friedland, G.J. Hawley, and R.A. Gregory, "Red Spruce (*Picea Rubens Sarg.*) foliar chemistry in Northern Vermont and New York, USA" *Plant and Soil* Vol. 105: 189–193, 1988.

129. Johnson and Siccama, "Decline of Red Spruce."

130. G.H.M. Krause, *et al.*, "Forest Effects in West Germany," Proc. Symp. Air Pollution and the Productivity of the Forest," Izaak Walton League, pp. 297–332, 1983.

131. Volker Mohnen, "Exposure of Forests to Gaseous Air Pollutants and Clouds," Report submitted to EPA, June 1987, Section 6.

132. Bruck, "Forest Decline in the Southeastern United States."

133. Huettl, "'New Type' of Forest Damages."

134. Ebasco Services Incorporated, "Acid Deposition Studies," 1–28.

135. National Acid Precipitation Assessment Program, *Interim Assessment, The Causes and Effects of Acid Deposition*, Volume IV, pp. 7–42.

136. Huettl, "'New Type' of Forest Damages."

137. Huettl, "'New Type' of Forest Damages."

138. G.A. Schier, "Response of Red Spruce and Balsam Fir Seedlings to Aluminum Toxicity in Nutrient Solutions," *Canadian Journal of Forestry Research*, Vol. 15, No. 1, 1985, pp. 29–33.

139. R.S. Turner, "Areas Having Soil Characteristics That May Indicate Sensitivity to Acidic Deposition Under Alternative Forest Damage Hypotheses," Oak Ridge National Laboratory, ORNL/TM-9917, 1986.

140. Turner, "Areas Having Soil Characteristics That May Indicate Sensitivity to Acidic Deposition," pp. 6–7.

141. Turner, "Areas Having Soil Characteristics That May Indicate Sensitivity to Acidic Deposition," p. 10.

142. Turner, "Areas Having Soil Characteristics That May Indicate Sensitivity to Acidic Deposition," p. 10.

143. Turner, "Areas Having Soil Characteristics That May Indicate Sensitivity to Acidic Deposition," p. 61.

144. Bruck, "Forest Decline in the Southeastern United States."

145. Johnson and Siccama, "Decline of Red Spruce."

146. G.H. Tomlinson, "Die-Back of Red Spruce, Acid Deposition, and Changes in Soil Nutrient Status—A Review" *Effects of Accumulation of Air Pollutants in Forest Ecosystems*, 1983, B. Ulrich and J. Pankrath eds. Reidel Publishing Company.

147. Walter C. Shortle and Kevin T. Smith, "Aluminum-Induced Calcium Deficiency Syndrome in Declining Red Spruce," Science, Vol. 240, May 20, 1988, pp. 1017–1018.

148. A.J. Friedland, G.J. Hawley, R.A. Gregory, "Investigations of Nitrogen as a Possible Contributor to Red Spruce (*Picea Rubens* Sarg.) Decline" *Symposium: Effects of Air Pollutants on Forest Ecosystems, May 8–9, 1985*, Minneapolis, Minnesota, University of Minnesota Press, pp. 95–106.

149. Bruck, "Forest Decline in the Southeastern United States."

150. Friedland, *et al.*, "Investigations of Nitrogen as a Possible Contributor to Red Spruce (*Picea Rubens* Sarg.) Decline," pp. 95-106.

151. Friedland, *et al.*, "Investigations of Nitrogen as a Possible Contributor to Red Spruce (*Picea Rubens* Sarg.) Decline," pp. 95-106.

152. Friedland, *et al.*, "Investigations of Nitrogen as a Possible Contributor to Red Spruce (*Picea Rubens* Sarg.) Decline," pp. 95-106.

153. A.J. Friedland and J.J. Battles "Red Spruce (*Picea Rubens*, Sarg.) Decline in the Northwestern United States: Review and Recent Data from Whiteface Mountain," *Proceedings of the Workshop on Forest Decline and Reproduction: Regional and Global Consequences*, Krakow, Poland, March 23-27, 1987, L. Kairiukstis, S. Nilsson and A. Straszak (Eds.), 1987 IIASA, A-2361, Laxenburg, Austria.

154. Huettl, " 'New Type' of Forest Damages."

155. B. Prinz "Causes of Forest Damage in Europe," 33.

156. Huettl, " 'New Type' of Forest Damages."

157. Prinz "Causes of Forest Damage in Europe." 32; Ebasco Services Incorporated, "Acid Deposition Studies," 1-29.

158. Prinz "Causes of Forest Damage in Europe," 33.

159. G.H. Tomlinson "Nutrient Deficiencies and Forest Decline," paper presented at Canadian Pulp and Paper Association Annual Meeting, Montreal, January 29, 1986.

160. Prinz "Causes of Forest Damage in Europe," 35.

161. Heck "Assessment of Crop Losses,"; National Acid Precipitation Assessment Program, *Interim Assessment, The Causes and Effects of Acid Deposition*, Volume IV, pp. 6-2 ff.

162. Heck "Assessment of Crop Losses."

163. U.S. Environmental Protection Agency, *Air Quality Criteria for Ozone and Other Photochemical Oxidants*, Volume III of V pp. 6-246.

164. Heck, "Assessment of Crop Losses."

165. P.D. Moskowitz, *et al.* "Effects of Acid Deposition on Agricultural Production," Biomedical and Environmental Assessment Division, Brookhaven National Laboratory, BNL-51889, September 1985, p. 1.

166. Heck "Assessment of Crop Losses."

167. U.S. Environmental Protection Agency, *Air Quality Criteria for Ozone and Other Photochemical Oxidants*, Volume III of V pp. 6-224.

168. Heck, "Assessment of Crop Losses."

169. U.S. Environmental Protection Agency, *Air Quality Criteria for Ozone and Other Photochemical Oxidants*, Volume III of V pp. 6-222.

170. U.S. Environmental Protection Agency, *Air Quality Criteria for Ozone and Other Photochemical Oxidants*, Volume III of V pp. 6-222.

171. U.S. Environmental Protection Agency, *Air Quality Criteria for Ozone and Other Photochemical Oxidants*, Volume III of V pp. 6-225.

172. U.S. Environmental Protection Agency, *Air Quality Criteria for Ozone and Other Photochemical Oxidants*, Volume III of V pp. 6-227.

173. U.S. Environmental Protection Agency, *Air Quality Criteria for Ozone and Other Photochemical Oxidants, Volume III of V* pp. 6-231.

174. U.S. Environmental Protection Agency, *Air Quality Criteria for Ozone and Other Photochemical Oxidants, Volume III of V* pp. 6-236.

175. Walter W. Heck, *et al.* "A Reassessment of Crop Loss From Ozone," *Environ. Sci. Technol.*, 17 (December, 1983): 579A.

176. Heck "Assessment of Crop Losses."

177. National Acid Precipitation Assessment Program, *Interim Assessment, The Causes and Effects of Acid Deposition*, Volume IV, pp. 6-44.

178. U.S. Environmental Protection Agency, *Air Quality Criteria for Ozone and Other Photochemical Oxidants, Volume III of V* pp. 6-252.

179. R.J. Kopp, W.J. Vaughan, M. Hazilla "Agricultural Sector Benefits Analysis for Ozone: Methods Evaluation and Demonstration," U.S. EPA, Office of Air Quality Planning and Standards, EPA-450/5-84-003 (Springfield, VA: NTIS, 1984, PB85-119477/XAB).

180. R.M. Adams, S.A. Hamilton, B.A. McCarl "The Economic Effects of Ozone on Agriculture," U.S. EPA, EPA-600/3-84-090 (Springfield, VA: NTIS, 1984, PB85-168441/XAB).

181. National Acid Precipitation Assessment Program, *Interim Assessment, The Causes and Effects of Acidic Deposition*, Vol. II, Emissions and Control, (Washington: U.S. Government Printing Office, 1987) pp 1-16 ff.

182. National Acid Precipitation Assessment Program, *Interim Assessment, The Causes and Effects of Acidic Deposition*, Vol. II, pp 1-42.

183. National Acid Precipitation Assessment Program, *Interim Assessment, The Causes and Effects of Acidic Deposition*, Vol. II, pp 1-17.

184. U.S. Environmental Protection Agency, *National Air Quality and Emissions Trends Report, 1985*, EPA-450/4-87-001 (Washington DC: U.S. Government Printing Office, 1987) 3-29.

185. National Acid Precipitation Assessment Program, *Interim Assessment, The Causes and Effects of Acidic Deposition*, Vol. II, pp. 1-45.

186. National Acid Precipitation Assessment Program, *Interim Assessment, The Causes and Effects of Acidic Deposition*, Vol. II, pp. 1-64.

187. U.S. Environmental Protection Agency, *Air Quality Criteria for Ozone and Other Photochemical Oxidants, Volume II*, EPA/600/8-84/020bF, (Washington: U.S. Government Printing Office, 1986) p 2-2; National Acid Precipitation Assessment Program, *Interim Assessment, The Causes and Effects of Acidic Deposition*, Vol. III, Atmospheric Processes, (Washington: U.S. Government Printing Office, 1987) pp. 4-7.

188. H.B. Singh, "Reactive nitrogen in the troposphere," *Environ. Sci. Technol.* 21 (April 1987): 320.

189. National Acid Precipitation Assessment Program, *Interim Assessment, The Causes and Effects of Acidic Deposition*, Vol. III, pp. 4-4.

190. J. Chamberlain, *et al. Acid Deposition*, JASON, The MITRE Corporation, JSR-83-301, 1985, page 2-7.

191. National Acid Precipitation Assessment Program, *Interim Assessment, The Causes and Effects of Acidic Deposition,* Vol. III, pp. 4–16.

192. National Research Council, National Academy of Sciences, *Acid Deposition, Atmospheric Processes in Eastern North America,* (Washington: National Academy Press, 1983), 40.

193. National Acid Precipitation Assessment Program, *Interim Assessment, The Causes and Effects of Acidic Deposition,* Vol. II, pp. 1–63.

194. National Acid Precipitation Assessment Program, *Interim Assessment, The Causes and Effects of Acidic Deposition,* Vol. III, pp. 5–65.

195. National Acid Precipitation Assessment Program, *Interim Assessment, The Causes and Effects of Acidic Deposition,* Vol. III, pp. 4–55.

196. National Acid Precipitation Assessment Program, *Interim Assessment, The Causes and Effects of Acidic Deposition,* Vol. III, pp. 4–71.

197. Bruck, "Forest Decline in the Southeastern United States."

198. V.K. Saxena, et al. "Monitoring the Chemical Climate of the Mt. Mitchell State Park."

199. U.S. Environmental Protection Agency, *Air Quality Criteria for Ozone and Other Photochemical Oxidants, Volume II,* pp. 3–68 ff.

200. Miller, "Concept of Forest Decline in Relation to Western Forests."

201. M. Trainer, et al. "Models and observations of the impact of natural hydrocarbons on rural ozone," *Nature* 329 (October 22, 1987): 705.

202. J.E. Pinkerton and A.S. Lefohn, "The Characterization of Ozone Data for Sites Located in Forested Areas of the Eastern United States," *Journal of the Air Pollution Control Association,* Vol. 37, No. 9 (September, 1987): 1005–1010.

203. U.S. Environmental Protection Agency, *Air Quality Criteria for Ozone and Other Photochemical Oxidants, Volume II of V* pp. 5–13 ff.

204. V.A. Mohnen, "Exposure of Forests to Gaseous Air Pollutants and Clouds," draft report to EPA of the Mountain Cloud Chemistry Project (MCCP), Contract Number 813-934010, June 1987, pp. 5–1 ff.

205. J.E. Pinkerton and A.S. Lefohn, "The Characterization of Ozone Data," 1005–1010.

206. Heck "Assessment of Crop Losses."

207. For a discussion of these methods see L.G. Chestnut and R.D. Rowe, "Economic Measures of the Impacts of Air Pollution on Health and Visibility," in forthcoming WRI book on Multiple Air Pollutants, 1989.

208. A.J. Krupnick, "Benefit Estimation and Environmental Policy: Setting the NAAQS for Photochemical Oxidants," Paper presented at the American Economic Association Meetings, New Orleans, LA, Dec. 1986.

209. Office of Air, Noise, and Radiation, Research Triangle Park, NC, U.S. Environmental Protection Agency, "Regulatory Impact Analysis on the National Ambient Air Quality Standards for Sulfur Oxides (Sulfur Dioxide)," Draft, May 1987.

210. Rowe et al., "The Benefits of Air Pollution Control in California," Prepared for the Air Resources Board, Sacramento, CA, 1986.

211. National Coal Association, "Reduction in Sulfur Dioxide Emissions at Coal-Fired Electric Utilities, The Trend Continues," Washington, D.C., 1987.

212. Michael Shepard, "Coal Technologies for a New Age," EPRI Journal, January/February 1988, p. 16.

213. National Acid Precipitation Assessment Program, *Interim Assessment, The Causes and Effects of Acid Deposition*, Volume II, pp. 3-9.

214. National Acid Precipitation Assessment Program, *Interim Assessment, The Causes and Effects of Acid Deposition*, Volume II, pp. 3-9.

215. National Acid Precipitation Assessment Program, *Interim Assessment, The Causes and Effects of Acid Deposition*, Volume II, pp. 3-26.

216. National Acid Precipitation Assessment Program, *Interim Assessment, The Causes and Effects of Acid Deposition*, Volume II, pp. 3-12.

217. The greenhouse problem derives from the buildup in the lower atmosphere of largely man-made gases (carbon dioxide, methane, chloroflourocarbons, ozone ("bad ozone"), and nitrous oxide). These gases are present in trace amounts in the atmosphere and are trapping an increasing fraction of the earth's escaping heat, threatening a global rise in the earth's temperature and long-term climate change. The depletion of the ozone layer ("good ozone") in the upper atmosphere (stratosphere) is primarily the result of chloroflourocarbons and nitrous oxide and could lead to serious ecological and health effects as well as enhanced greenhouse impacts.

218. National Research Council, National Academy of Sciences, "Acid Deposition, Atmospheric Processes in Eastern North America," Washington, D.C. 1983.

219. Office of Technology Assessment, U.S. Congress, "Acid Rain and Transported Air Pollutants, Implications for Public Policy," U.S. Government Printing Office, Washington, D.C., 1984, p. 53.

220. W.D. Baasel, "Capital and Operating Costs of Wet Scrubbers Installed on Coal-Fired Utilities Impacting the East Coast," *Journal of the Air Pollution Control Association*, Vol. 38, No. 3, March 1988 pp. 327-332

221. National Acid Precipitation Assessment Program, *Interim Assessment, The Causes and Effects of Acidic Deposition*, Vol. II, pp. 2-42.

222. H.S. Geller et al. "Acid Rain and Electricity Conservation," American Council for an Energy Efficiency Economy, Washington, D.C. June 1987.

223. Michael P. Walsh, "Pollution on Wheels," A Report to the American Lung Association, Washington, D.C., February 11, 1988.

224. The new standards for hydrocarbons would cut emissions by 40 percent relative to the current standard. NO_x emissions for autos would be cut by 60 percent.

225. Michael P. Walsh, "Pollution on Wheels," A Report to the American Lung Association, Washington, D.C., February 11, 1988, Figure 17.

226. Personal Communication, Michael P. Walsh, July, 1988.

227. M.A. DeLuchi, R.A. Johnston, D. Sperling "Transportation Fuels and the Greenhouse Effect," UER-180, Division of Environmental Studies, Univ. of California, Davis, December 1987.

228. U.S. Department of Energy, "Energy Security" A Report to the President of the United States," March 1987, p 167.

Appendix

Authors of Background Studies

Prof. Robert I. Bruck
Department of Plant Pathology
North Carolina State University
Raleigh, North Carolina

Ms. Lauraine G. Chestnut
RCG/Hagler, Bailly, Inc.
Boulder, Colorado

Prof. Walter W. Heck
Department of Botany, and
Chairman, Research Committee
National Crop Loss Assessment Network
North Carolina State University
Raleigh, North Carolina

Prof. Reinhard F. Huettl
Institute for Soil Science and Forest Nutrition
Albert-Ludwig University,
Freiburg, Federal Republic of Germany

Professor Arthur H. Johnson
Department of Geology
University of Pennsylvania
Philadelphia, Pennsylvania

Dr. Paul Miller
USDA Forest Service
Pacific Southwest Forest and
Range Experiment Station,
Riverside, California

Dr. Robert D. Rowe
RCG/Hagler, Bailly, Inc.
Boulder, Colorado

Prof. Thomas G. Siccama
Yale School of Forestry and
Environmental Studies
Yale University
New Haven, Connecticut

Advisory Panel

Mr. William Becker
Executive Director
State and Territorial Air Pollution Program
  Administrators
Washington, D.C.

Prof. F. Herbert Bormann
School of Forestry and Environmental Studies
Yale University
New Haven, Connecticut

Prof. Ellis B. Cowling
School of Forest Resources
North Carolina State University
Raleigh, North Carolina

Ms. Katherine Y. Cudlipp
Chief Counsel
Committee on Environment and Public Works
U.S. Senate
Washington, D.C.

Mr. David Hawkins
Senior Attorney
Natural Resources Defense Council
Washington, D.C.

Prof. Walter W. Heck
Department of Botany, and
Chairman, Research Committee
National Crop Loss Assessment Network
North Carolina State University
Raleigh, North Carolina

Mr. Thomas Jorling
Commissioner
Department of Environmental Conservation
Albany, New York

Dr. Richard Klimisch
Executive Director
Environmental Activities Staff
General Motors
Warren, Michigan

Dr. William H. Lawrence
Senior Research Advisor
Weyerhaeuser Corporation
Centralia, Washington

Dr. Orie Loucks
Holcomb Research Institute
Butler University
Indianapolis, Indiana

Mr. James Lyons
Staff Assistant
Committee on Agriculture
U.S. House of Representatives
Washington, D.C.

Dr. Ralph Perhac
Director of the Environmental Science Dept.
Electric Power Research Institute
Palo Alto, California

Dr. Benjamin B. Stout
National Council of the Paper Industry for Air
    and Stream Improvement
New York, New York

WRI PUBLICATIONS ORDER FORM

ORDER NO.	TITLE	QTY	TOTAL $
756	*Ill Winds: Airborne Pollution's Toll on Trees and Crops* by James J. MacKenzie and Mohamed T. El-Ashry, 1988, $10.00		
750	*The Shrinking Planet: U.S. Information Technology and Sustainable Development* by John Elkington and Jonathan Shopley, 1988, $10.00		
787	*The Forest for the Trees? Government Policies and the Misuse of Forest Resources* by Robert Repetto, 1988, $10.00		
788	*Money to Burn? The High Costs of Energy Subsidies* by Mark Kosmo, 1987, $10.00		
709	*Energy for a Sustainable World* by José Goldemberg, Thomas B. Johansson, Amulya K.N. Reddy and Robert H. Williams, 1987, $10.00		
785	*Energy for Development* by José Goldemberg, Thomas B. Johansson, Amulya K.N. Reddy and Robert H. Williams, 1987, $10.00		
718	*Multinational Corporations, Environment, and the Third World: Business Matters* edited by Charles S. Pearson, 1987, $17.95 (paperback)		
786	*Not Far Afield: U.S. Interests and The Global Environment* by Norman Myers, 1987, $10.00		
789	*To Feed The Earth: Agroecology for Sustainable Development* by Michael J. Dover and Lee M. Talbot, 1987, $10.00		
710	*A Matter of Degrees: The Potential for Controlling the Greenhouse Effect* by Irving M. Mintzer, 1987, $10.00		
792	*Skimming the Water: Rent-Seeking and the Performance of Public Irrigation Systems* by Robert Repetto, 1986, $10.00		
783	*The Sky Is the Limit: Strategies for Protecting the Ozone Layer* by Alan S. Miller and Irving M. Mintzer, 1986, $10.00		
781	*Double Dividends? U.S. Biotechnology and Third World Development* by John Elkington, 1986, $10.00		
719	*Bordering on Trouble: Resources and Politics in Latin America* edited by Andrew Maguire and Janet Welsh Brown, 1986, $14.95 (paperback)		
784	*Troubled Waters: New Policies for Managing Water in the American West* by Mohamed T. El-Ashry and Diana C. Gibbons, 1986, $10.00.		
712	*Growing Power: Bioenergy for Development and Industry* by Alan S. Miller, Irving M. Mintzer, and Sara H. Hoagland, 1986, $10.00.		
725	*Down to Business: Multinational Corporations, the Environment, and Development* by Charles S. Pearson, 1985, $10.00		
723	*The Global Possible: Resources, Development, and the New Century* edited by Robert Repetto, 1986, $14.95 (paperback)		
732	*World Enough and Time: Successful Strategies for Resource Management* by Robert Repetto, 1986, $6.95 (paperback)		
724	*Getting Tough: Public Policy and the Management of Pesticide Resistance* by Michael Dover and Brian Croft, 1984, $10.00		
714	*Field Duty: U.S. Farmworkers and Pesticide Safety* by Robert F. Wasserstrom and Richard Wiles, 1985, $10.00		
716	*A Better Mousetrap: Improving Pest Management for Agriculture* by Michael J. Dover, 1985, $10.00		
717	*The American West's Acid Rain Test* by Philip Roth, Charles Blanchard, John Harte, Harvey Michaels, and Mohamed El-Ashry, 1985, $10.00		
715	*Paying the Price: Pesticide Subsidies in Developing Countries* by Robert Repetto, 1985, $10.00		
776	*The World Bank and Agricultural Development: An Insider's View* by Montague Yudelman, 1985, $10.00		
731	*Tropical Forests: A Call for Action*, 1985 by WRI, The World Bank and UNDP, $12.50		
726	*Helping Developing Countries Help Themselves: Toward a Congressional Agenda for Improved Resource and Environmental Management in the Third World* by Lee M. Talbot, 1985, $10.00		
780	*World Resources 1987*, $16.95 (paperback)		
	SUBTOTAL		
	POSTAGE & HANDLING ($2.00 first title, 50¢ for each additional)		
	TOTAL DUE		

Name

Street Address

City/State					Postal Code/Count

Please send check or money order (U.S. dollars only) to WRI Publications, P.O. Box 620, Holmes, PA 19043-0620, U.S.A.

CLIFTON M. MILLER LIBRARY
WASHINGTON COLLEGE
CHESTERTOWN, MARYLAND 21620